THE
EDGE
OF
DARKNESS

Novels by Mary Ellen Chase

THE
EDGE
OF
DARKNESS

Mary Ellen Chase

NEW YORK

W · W · NORTON & COMPANY · INC.

Library of Congress Catalog Card No. 57-10637

PRINTED IN THE UNITED STATES OF AMERICA
FOR THE PUBLISHERS BY THE VAIL-BALLOU PRESS

FOR

Newton Kimball Chase

AND

Elizabeth Kilbourn Chase

THIS STORY is set in no definite or discernible place, but instead in any one of many small and isolated fishing communities on a coast long familiar to me. The names chosen for the characters are names common to the coast of Maine in its more than two centuries of seafaring. If any of them are possessed by actual persons, chance and not intention is responsible.

Mary Ellen Chase

NORTHAMPTON, MASSACHUSETTS
APRIL, 1957

Contents

✧ ─────────────────────────────

PART ONE

Sarah Holt

❖ ───

Sarah Holt

❖ ─────────────────────────────

L UCY NORTON stood beside old Mrs. Holt's
coffin and looked down at the woman who
had been her neighbor for thirty years.
This was clearly her last call on Sarah Holt; and she felt
annoyed and even guilty that she could not seem to keep
her mind securely anchored to the still, sharply-lined face
against the white satin pillow. No sooner had she noticed
how fine and straight the nose of the dead woman was, how
her hands with their long fingers did not reveal the hard
work they had done for half a century, and how, after all
her ninety years, she did not look shrunken or wasted as
most of the dead look, than her thoughts and even her
suspicions flew outside to Thaddeus Holt, who was pacing
restlessly up and down the high beach. She wondered what
was taking place in his mind as he walked to and fro, and
whether, because of his loneliness and remorse, he would be
able to fulfill his mother's last odd request. And when she
had pulled her thoughts back once more and marked with
an inward smile how nice and fresh the old-fashioned white

fichu looked against Sarah Holt's black dress, they began to
wander away again, not to all the many times she had
seen the fichu or washed and ironed it for Sarah over all
the years, but, instead, to Joel's blue suit spread out upon
the bed at home. Had she succeeded in getting the shine
off it by numberless steamings? And had Joel, who was apt
to be forgetful even on ordinary days, remembered to
stock up heavily with frankfurters on his trip to town at
dawn? There was bound to be a run on them at the store
after the funeral, after two days, in fact, of little cooking in
the face of so much to talk about. Then her mind slipped
away to the tide, which had already turned and now was
beginning to right the skiffs and the dories in the cove; to
the dim outlines of the farthest fishing-boats coming in
after their early morning's haul; and at last to the time itself, for it was getting on for eleven as she could see from
the tall old clock between the two front windows and
just behind the coffin, and she had more to do than she could
possibly manage to do before the funeral at two that afternoon.

The clock with its menacing hands startled her into fresh
and painful realization that this last visit to Sarah Holt was
not in the least what she had planned and wanted it to be;
and now, in spite of its warning that time was hastening on,
it perversely reminded her of itself. She recalled how Captain Holt had bought it in London as a present for his bride
on her first voyage with him in those old days of sail. They
had brought it home, swathed in canvas and lying on its
long back in the hold of his ship, securely anchored there
against the pitching and the rolling. One night, when the
sailors had been shifting cargo in a wild mid-Atlantic

storm, it had suddenly begun to strike, down in the dim bowels of the ship, its reverberating, somber tones sounding above the ominous creaking of planks, bolts, and gear, and quickening all hearts with terror.

Whatever had been Captain Holt's long name? He had been thirty years older than Sarah, who had married him when she was only a girl of eighteen. Now she had it: Thomas Jefferson Alexander Hamilton Holt. For he had actually been born when those names still meant a great deal in history, and he had died fully half a century ago. He had walked about these rooms, for the house had belonged to his family, sat in the chairs, scanned the weather, deplored the rotting piers and docks of a hundred Maine harbors, and cursed the death of sailing-ships before the inroads of steam. He had brought up his only son, Thaddeus, to dream long dreams which time and change had made impossible of realization. Thaddeus at sixty was obsessed now by dreams of quite another sort. Lucy could hear his heavy feet crunching through the beach gravel in the incredible stillness of this late September day. She could see his long, thick fingers nervously writhing behind his back. She felt herself unwillingly torn by pity for him, although what many people felt, and perhaps rightly, was contempt and scorn.

Now her errant, distressing mind, skipping all over the place in spite of her guilt and desire, recalled something she had once read in a story lent her by Sarah Holt. Lucy had not read many books before she had known Sarah, and at first she had found some of them hard going enough. In this story some character in a long and rather puzzling dialogue had said that one was never able to catch and to hold the essence, the real meaning of anything at all; that just as you thought you had it, it slipped from your grasp and left you searching and lonely for what it might have told you about life and what things really meant. This was a new thought to Lucy, explaining the confusion she so often felt and consoling her in that others had felt the same. It returned now to reproach her as she stood by Sarah's coffin in this room which had given her solace, understanding, and courage for thirty years.

Then, as if to strengthen the truth of the first, and perhaps to afford uneasy absolution also, there came another story from yet another book. This story had to do with a young priest, who was tormented by guilt because, just as he was about to consecrate the bread and the wine at the altar, he could not hold his mind upon his sacred act. It kept slipping from him, running hither and yon, even mockingly roaming away to boyhood lusts and to sinful desires which were shameful in his dedicated life. Lucy had suffered for this young man at odd moments ever since she had read about him; and she understood his anguish more completely as she now strove with her own.

For she did not want in this last hour with Sarah Holt merely to remember things about her. She would have plenty of time for that. With the autumn just around the corner and the winter quite too close at hand, she would have hours on end to remember a thousand things. With Joel away for provisions and trade slack at the store, she could sit behind the counter, watch the flooding or the ebbing tide, and recall at will any number of pictures, strange sights, and memorable conversations: how she had first seen Sarah Holt pacing the beach and scanning the sea; how she had first heard from her what this strip of coastline had been like when Sarah was young; how she had learned, through Sarah, to look upon it as it was to-day. All these things she could remember, with countless other intimate, funny, congenial exchanges between them; but that was not enough.

Memories are not realities. They are as different one from the other as is the path of a shooting star from a fixed constellation, or as the moving tide from the mysteri-ous instant of its complete fullness. Memories heal, or amuse, or comfort, or even nurture; but they cannot alone fortify or make invulnerable. They can enliven; but they cannot sustain. Only the radiance of sudden insight or of vision can work that miracle.

What Lucy Norton wanted, what she had come for and foolishly had expected to find here alone with Sarah, until her mind had defeated her, was the seizure, if only for a moment, of the meaning of Sarah Holt's long, hard, triumphant life. If she could only grasp that meaning, ap-prehend it for the briefest instant, lose herself and her confusion in the consciousness of it, she would want for nothing. She could face the winter after a lean fishing season; weather the northeast gales of the late autumn with their toll of weir-posts and lobster traps; endure wobbling credit at the store, Joel's anxious concern, disgruntled, churlish, frightened men and women, and ill-fed children.

She became all at once aware of a new silence in the room, more heavy and pervasive than that stillness which death always brings to a house. She could almost hear, see, and smell this silence. It was as though all the thoughts of the dead woman had come trooping back to weave in and out, among the chairs and chests and tables, over and around the pictures on the walls, like elusive fragrances, or wraiths of mist, or the low hum of bees—all her hopes and sorrows, small comforts, and lingering realities. These would be gone by tomorrow, Lucy knew, borne far away, and only the shabby details of the house itself would remain when she came over, as of course she would, to set things in order again for Thaddeus.

The panic of this knowledge, together with the enveloping silence, drew at long last a thick curtain over all the nagging anxieties which had been claiming her thoughts: Joel's suit, the frankfurters, the incoming tide, the time itself, even over the words of the books which had added to her remorse. She no longer heard the hurtling of the beach gravel from Thaddeus' heavy feet or saw his restless hands. Now the few minutes left to her were embracing thirty years, making a circumference of them, a great circle, which started on the day when she and Joel had come to this point of land facing the open Atlantic and which now had closed with Sarah Holt's death. She stood within that shining circle, that circumference, just as though it were outlined in light upon the braided rug of the sitting-room floor. She was alone there for these brief moments which had become half a lifetime. She was not a woman at all, not Lucy Norton in a black-and-white checked gingham dress, with crooked, rimless spectacles over her near-sighted eyes, who with her husband ran the general store in the fishing village. She had lost all identity, all consciousness of herself. She had become blotted out, erased, lost, in what seemed to be succeeding waves of understanding, engulfing waves of surprise and wonder, of pity, hope, and faith, and one final overwhelming wave of gratitude.

She supposed that the echoing tones of the clock striking eleven brought her back to where she actually was, on the braided rug beside the coffin. The circle of light had gone. She was again in her checked gingham. Her hands and her feet, which had been so strangely absent, were here again, ready for all the things that awaited them. She no longer felt confused or distracted. She had at last got what she had come for.

2

Before she left the Holt house, which stood in a field above the sea and a quarter of a mile from the one road that led through the village, she went through the rooms to see that everything was in order for the funeral. First, the kitchen, which was back of the sitting-room. The children would sit here, on the low stools and little chairs which Thaddeus had made for them. Thaddeus was good with his hands, and between his spells of drinking he had at his mother's suggestion framed six or eight of these for the children of more than one generation to sit upon when they came to the house to see her, to listen to her stories, eat her cookies, and show her how well they could read. He had painted them in bright colors, yellow, blue, and red, and the children loved them. They would come to the funeral as would everybody else in the village, except old Daniel Thurston, who was sick again, the Randalls, who would not dare to come, and perhaps Drusilla West, about whom no one felt certain. They would sit there in the kitchen, all well-scrubbed, all shy and curious, and perhaps a bit frightened. It was this notion of their being frightened which made Lucy Norton arrange the chairs and stools in a small circle in the center of the clean kitchen floor. If they could sit there close together as they had so often sat, they might feel more at ease and at home.

Lucy always liked to see and handle these little stools and chairs. There was one red chair among them, which in an odd manner always made her want both to smile and to cry. Thaddeus had made its backboard longer than those of the others and had taken the pains to insert tiny, rounded spokes which connected the backboard to the back of the seat, and even to fashion small, rounded arms for it. It was queer, Lucy thought, looking at it, how most people when they made things just made things, no matter how carefully they might work; and then how someone else would come along who could fashion a chair, or a little lobster buoy, or a small boat so that it took upon itself both nature and being, and was able in an indefinable, yet inescapable way, to arouse pleasure, or pain, or even a kind of wisdom.

She left the kitchen and went into the parlor, which was opposite the sitting-room, across the hall, beyond the staircase. Sarah Holt had not fancied her parlor and had seldom sat there; but with extra people doubtless coming from the villages and towns along the main highway, not to mention the backwater folks, it, too, must be reckoned with like the bedroom behind it. Lucy now saw to the parlor, straightened the chairs, which Thaddeus, glad of something to do, had carried downstairs or brought in his truck from the houses of the neighbors, and made sure that all the window-shades were in exact line with one another.

The neighbors would, of course, sit with Thaddeus in

the sitting-room, there being no members of the family. She named them over, silently on her fingers although there was no need—eleven in all, twelve counting the doctor, who would be sure to come even with his fifty-mile drive and at the neglect of his afternoon patients. She did not include the Randalls, except perhaps for their little girl, who might be allowed to come with the other children if she begged hard enough, or who might come anyway if she were left alone as she often was. The sitting-room was much larger than the parlor, and the twelve chairs were carefully placed in the back of the room for all.

3

There was, of course, more than a bare possibility that Thaddeus' wife would come. One never could be sure about Nan Holt. That warped sense of obligation to finish what she had started twenty-five years ago when she had unwisely married Thaddeus was still strong within her. Sarah Holt had hammered against it for fifteen years until she had weakened it sufficiently to induce Nan to leave, to return to her school-teaching, to go her own way. But even Sarah had never completely demolished it. It was still alive, resolute and uneasy, torturing Nan with a sense of guilt, and, worse still, with the pain of her love for Thaddeus. For, incredible as it seemed, she still loved him.

Now and again, during the ten years since she had gone, she wrote to him, giving him news of their only son, Jeff, who at seventeen had run away from all that had made his boyhood wretched and was now doing well in the wheat fields of Kansas. She never enclosed one of Jeff's infrequent letters with her own to Thaddeus, partly because she knew that the sight of his handwriting would bring bitter regret and remorse to his father, mostly because Jeff had not yet become able to subdue his old resentments. He still railed against prying, contemptuous neighbors, humiliating scenes at home, and sudden silences when he came among the men framing their traps in the fish-houses in the winter; against cold dawns in his father's plunging boat or equally cold, wet nights when they seined the weir by lantern light; even against the stench of herring bait. He hoped he would never see any of it again, he wrote cruelly to his mother, who understood, even as she cried over his letters, that cruelty in one form begets cruelty in another. When he awoke in Kansas, he said, and saw the sun rising over illimitable acres of flat, rich land, the windmills slowly turning, and herds of cattle among groves of catalpa trees, he had never been so near happiness in all his life. He would like to see his grandmother again, and perhaps Sam Parker and the Nortons; but the cost was far too great to pay. As to his mother, there were, after all, as she must know, schools in Kansas and far better ones, too, than those in Maine.

As soon as Nan Holt had amassed six or eight of these letters, which happened once a year or so, and had carefully selected just the right details to relay to Thaddeus, she put them all in a long envelope, carefully addressed to "Mrs. Thomas J. Holt," and sent them home. She had no fear of doing this. If Thaddeus should be aware of their arrival, which was improbable, he would never ask his mother about them, not only because of his respectful fear of her, but because he still cherished, even in his worst hours, an odd and even chivalrous code of behavior. And since Sarah Holt had been the power behind Jeff's release also, it was only just and right that she should be given all the news there was of her grandson.

She had always shared Jeff's letters with Lucy Norton. After a packet of them had come, she read them aloud during an hour or two in the late afternoon when they sat together in the kitchen or the sitting-room, knitting bait-bags or trap-heads out of strong, rough twine. Kansas seemed to them both an unimaginable distance away. Nor could either conceive of such vast stretches of mere land.

"And I can't see why not," Sarah Holt said. "I've always been able to see most things in my mind, and God knows I've seen enough places in my life. But just land alone baffles me. All the books I've read about the West and all the pictures I've seen of it simply don't open my eyes. Now I come to think of it, no port we ever made had much open land around it. Perhaps that's the trouble. There were usually mountains behind them, like San Francisco and Rio and Marseilles; or muddy rice fields, like the far Eastern ports; or islands rising all around them, like the Mediterranean harbors. There was flat land in Holland, to be sure, but there was always water about in the canals. I just can't seem to imagine, Lucy, what it must be like to see horizons surrounding miles and miles of just dry land."

"I can't either," Lucy said, as she rose to put another stick of wood in the stove and set the kettle to boil for their tea.

"The nearest I can seem to come to it is the doldrums with the sea flat and still as far as you could see, not a breath of wind and no steerage-way at all. You always struck rock bottom in your mind after days on end in the doldrums. But perhaps flat plains don't affect one in that way; and if Jeff couldn't take to the sea like all the rest of us or even abide living by it, I'm glad he likes all that motionless land."

"There's plenty of wind in Kansas," Lucy said. "I've read often of awful cyclones there."

"There are cyclones everywhere," Sarah Holt said, "of one sort or another."

Lucy, fearful of the older woman's eyes, welcomed the rising hum of the tea-kettle.

"I remember now," Sarah Holt said, "that Jeff's grandfather knew about vast stretches of just land. He was fond of telling me how, when he shipped to Buenos Aires after a cargo of hides way back in the eighteen-forties before I was so much as born, he used to visit a cattle-raiser out on those Argentine plains. They rode horseback all day, he said, on and on with nothing to see but grass and sky and cattle. I remember he said that the grass there was almost a purple color and beautiful, waving in the wind. Some day I must write to Jeff about that. I don't want him to forget about his grandfather."

Lucy now pulled from the pocket of her gingham skirt the duster which she always carried about with her and wiped some real or imaginary specks of dust from the parlor tables, the white mantelpiece, and the Franklin stove beneath it. Then she perked up the flowers in their jars and vases. She was still uneasy about Nan Holt. She did not fear so much her coming to her mother-in-law's funeral as she dreaded the commotion it would cause and the inevitable talk it would make which she would have to meet, head on and single-handed. To be sure, she had met it in that way ten years ago and a hundred times since then; but there had always been Sarah Holt in the background, like some immovable granite boulder.

When she went into the bedroom to be sure there was no dust on the clothes-chest there and on the posts and headboard of the great old bed, she rehearsed to herself the part she had played for Sarah on the day that Nan had left Thaddeus for good, after several half-hearted attempts to do so.

"You tell them, Lucy," Sarah had said. "They'll never dare to say a word to me. Tell them in your own way. You're a born actor, and you can at least make them listen. Tell them that I drove Nan away from Thaddeus, which is the plain truth. Say that I didn't want her here and that I wouldn't let her stay. Thaddeus doesn't deserve her kind of love. He might have once, but he doesn't any longer. And if he wants just women, he can find them in a hundred places on this God-forsaken coast, though you needn't tell them that. She's too good for the rampage and ruckus of this house, and even Jeff is better off just with me. He won't be here much longer anyhow. I brought Thaddeus into this world, and he's mine to pay for. Tell them that I said just that."

Lucy said the words over to herself, realizing as she did so that additions and emphases of her own would be necessary in view of this change in circumstances. When she left the bedroom and went through the kitchen for a last moment in the sitting-room, she carried another chair with her to add to the twelve already there.

4

She was hardly prepared for the difference in the atmosphere of the sitting-room. She might have stepped into quite another room from that which she had left ten minutes before. When she had set the extra chair down, she moved to the center of the braided rug and stood there, wondering just wherein the difference lay. She was so sure that things were not as they had been that she found herself taking off her spectacles, breathing on them, polishing them with her handkerchief, and putting them on again. She had performed this silly gesture for years whenever her mind felt confused or uncertain, as though her clean lenses could help to straighten it out.

She was convinced that the air in the room had lost its heaviness, become lighter, more transparent. Perhaps it's just me, she thought, that I'm not anxious or flurried any more. This explanation did not satisfy her. The room was still, but it was not silent any longer, only tranquil and quiet. It was like that clear, translucent light at the close of a day overcast and darkened by fog, which, when the fog has rolled seaward before it, reveals the distinct, familiar outlines of shore and islands. The sitting-room was now as bright as the close of such a day.

Lucy smiled there, in the middle of the braided rug, for she fancied that she knew what had wrought this transformation. There were no restless thoughts in the corners and about the walls, no more sorrows, no desperate decisions fraught with pain, no hopes, regrets, or memories. Sarah Holt's soul, or spirit, or whatever it was that had given her life, had gone. She had taken it with her, not waiting for Shag Island where they would carry her body in the late afternoon, but already to some undiscovered new place of being, in an unknown time and space. Perhaps her soul had been still hovering about the old house, reluctant to leave the things it had always known. Perhaps it had been lingering there for someone to comprehend the meaning of its sojourn upon this particular spot of earth. This notion was as comforting as it was exciting, and Lucy smiled again.

She was filled now with an odd desire to say a prayer for Sarah Holt's soul, winging its way somewhere beyond that far, clear horizon where sky and open sea met. Crazy as such an impulse seemed and close as she would always keep it within herself, she wanted to kneel here in this now cheerful room and say just something. She remembered some words which she had heard somewhere years back, or perhaps read, and which had always echoed pleasantly in her mind: *May the souls of the faithful departed, through the mercies of God, rest in peace.* Those would do. Her cheeks grew flushed as she thought of the many times she had said them over to herself at coast or island funerals, often in small family burying-grounds, as the wind from the sea whipped the clothes of the few people standing about and an awful loneliness increased the alienation of one from another.

She said the words now in a half whisper for Sarah Holt's soul. She could not quite bring herself to kneel down on the rug, much as she longed to do so. To kneel would be in defiance of her upbringing and tradition; and she might feel unreal and even dishonest just as she had

(28)

felt the night before last when the doctor had dropped her at the store after their vigil together in Sarah Holt's bedroom.

She had gone into the store just before midnight to find, as she had expected, all the fishermen there, sitting about in uneasy silence, smoking their pipes with Joel, waiting for her news. She answered their voiceless, common question with the doctor's words. She was quite safe in seizing upon them as her own, for the doctor had swayed on in the swirling fog to see old Daniel Thurston a mile farther on up the hill.

"She's just gone," she said to the small group of tired men. "She marks the end of an era on this coast."

If the fishermen were impressed, they did not show it. They knocked out their pipes against the stove or on the big knuckles of their hands and went home for four hours' sleep before the dawn sent them stumbling to their dories. Only Sam Parker said goodnight at the door.

But when she and Joel were getting ready for bed in their rooms above the store, he had stared at her with admiration and puzzled wonder.

"No one else would ever put it that way," he said. "You always know the right way to say things, Lucy. It beats me how you always think of just the proper words. She is the end of old days on this coast, that's sure, but whoever would have thought to call them an *era?* I'd never so much as considered that word till I heard you say it."

When they were in bed, Lucy lay close to Joel's broad back, crying a little in the wet darkness which crept through the window. Joel had fallen asleep almost at once.

She would tell him some day, she resolved, that she also would never have thought of an *era* until the doctor called it that.

5

The doctor had not reached the Holt place until after ten o'clock the night Sarah Holt died. When he came through the back door and stopped for a moment in the hall to shake his drenched overcoat before he hung it on a peg there, he told Lucy, who was sitting by Sarah's bed, that the fog was the worst he had ever seen. It had taken him almost three hours to drive fifty miles.

"My lights are about as much good as two kitchen matches," he said, "and my windshield's no better than a sponge. The main road's bad enough, God knows, but when you turn off it for this cove where you folks live, you're driving through Hell itself."

He went into the kitchen to wash his hands. Thaddeus was sprawled across the table there, his head with its heavy white hair resting on his arms. He wore a gray flannel shirt and looked well-shaven and clean. An empty bottle and a tumbler were on the table. The doctor moved these to the shelf beyond the sink.

"How long has he been at it this time?" he asked Lucy through the open bedroom door.

"All day, I guess," Lucy said. "He's been wandering along the shore. He's kept away from the house. The men couldn't do anything with him when they got in. They was late with all this fog. I got his supper for him, but he wouldn't eat. He hasn't said a word since eight o'clock. Then, when I told him 'twas shameful to carry on so at a

(30)

time like this, he just said he didn't mean to touch it for the next two days. But you never know."

"Poor devil!" the doctor said. He closed the kitchen door before he came into the bedroom.

Lucy sat by the head of the bed under the Aladdin lamp on the table. She had Sarah Holt's mending-basket on her lap, and she was darning some socks between her quick glances at the woman lying quietly against the clean pillows. There was no sound in the room, or, if there was, from the low hiss of the blue mantle of the lamp or from the short, quick breaths from Sarah Holt's lips, it was lost in the uproar outside.

The tide was only just past the half, but before a high southwest wind it was crashing up the beach, each receding roller dragging down a grinding weight of shale and gravel. In the brief moments between the flooding and the ebbing waves, they heard the slap of angry water against rocking dories and fishing-boats; the banging and shifting of unsteady gear, buoys and traps piled against the shed and barn; the clank of oarlocks and anchor chains; and, nearer at hand, the ceaseless dripping of fog-filled lilac bushes and swaying spruces.

Before the doctor sat down in the chair which Lucy had placed for him, he loosened the high neck of Sarah Holt's nightgown and listened through his stethoscope to her chest and heart. Then he clasped her wrist, thin and blue against the white bedspread, with his hand.

"It won't be long now," he said.

"Shall I try to rouse Thaddeus?"

"I don't see the sense. I don't think you could anyway, and if she comes to at all, which I doubt, she won't fancy

(31)

seeing him like that. He's safe enough there. He won't come out of it till morning."

The noise outside increased as the tide rose higher. They seemed in some small, frail shell, Lucy thought, which the sea did not cease to toss about and torment.

"Is she as quiet as she looks?"

"Yes, I'm sure she is. If I wasn't sure, I'd do something. There isn't anything to do now except just wait."

He looked at Lucy. She had put the mending-basket on the table beyond the lamp. She had taken off her spectacles and was breathing against them. Now she polished them carefully with her handkerchief before she put them back upon her nose.

"How many times have you had to see this, Lucy?"

"Numbers," Lucy said.

They did not speak for what seemed a long time to Lucy, but she felt no sense of strain between them. She had hoped all along that it would end like this, with just her and the doctor. She forgot her outrage against Thaddeus. She was glad he was not here.

There was a sound of splintering wood outside, in a brief pause between the crashing of the waves upon the beach. Traps, Lucy thought. He should have taken them inside in a blow like this.

The doctor was talking again.

"But not the same as this time," he was saying. "You and I will never see anything just like this again. I thought about it coming out whenever I wasn't cursing that terrible road or finding myself in a ditch. I kept thinking how old Mrs. Holt's going marks the end of an era on this coast. She was ninety last month, she told me. That means she was

(32)

born in the beginning of the clipper-ship days. As long as she was alive, there was something of the past about, but that goes out with her."

"She sailed everywhere with her husband when she was young," Lucy said. "For twenty years until the steam began to take over. But you know all about that as well as me."

"In those days when she was born," the doctor said, "this coast was closer to India and China than it was to the Middle West, and it kept on being so for thirty years longer. Even the wretched town I live in built ships, ten to twenty every year, and sent men all over the world in them. People thought of other things then than just packing herring and canning blueberries."

"They built ships even here, too," Lucy said. "Up the Tidal River there were great docks and yards. You can still see up there the ruins of the ways that the ships slid down at their launchings. You can still see them right here, for that matter, over on the headland just beyond Dan Thurston's place where the water is deep and sheltered. There are old rotting piers and timbers on Shag Island, too, where she was born and where she wants to be buried. There hasn't been a soul on Shag Island for twenty years except hunters in November. When we first came, there were a few fishermen, but they've left for the mainland. It's queer to think of that island any different than what it is now."

"Everything was different then, people as well as places," the doctor said. "You couldn't sail all over the world and stay shut up in your own mind. And even if you didn't sail, you knew those who did, and got familiar with strange ports and countries and the way people lived in them. She never said much to me about her life at sea in the five years I've been coming out here to see her, off and on, but I suppose she did to you."

"Not so much lately," Lucy said. "But you always saw it all back there in her thoughts. She had plenty enough here to trouble her, and I think those old days kind of stayed her heart."

"That's another thing the sea did for those who sailed it. It put iron into folks so that they could take things the way she did and not crumple up under them."

Lucy saw now that his fingers were moving upon Sarah Holt's wrist, pressing it here and there, and that his eyes, even as he talked, never left her face.

"I brought her some of my roses a little while ago. I remember she told me then of the roses she'd seen all over the world. She named a lot of places, France and the Azores and England. She said, though, that she'd never seen nicer roses anywhere than mine."

"How are your roses doing now?" Lucy asked politely.

"Fine. Better every year. There's some still in blossom even this late. They seem to fancy baked clay and gravel and ground-up clam shells. If it weren't for those roses, I swear I'd have skedaddled from this coast long ago."

"I often wonder why you stay," Lucy said. "For the matter of that, I've wondered more than a few times why any of us do."

The doctor drew some gauze from his bag which he had placed at the foot of the bed, moistened it in a tumbler of water on the table, and wiped Sarah Holt's lips.

"When you think up the answer to your question, let me know," he said. "It will probably be the same as mine and will save me a lot of time."

Lucy moved in her chair as though she were about to leave it.

"Would you think it foolish of me to light the candles there on the mantelpiece?" she asked. "Joel got fresh ones for her just two days ago. She always liked them lighted there, and those big old silver holders came from London at the same time as the clock in the sitting-room. I polished them for her just this morning, and I think she noticed them. If she did just open her eyes even for a minute, she'd like seeing them."

"I think it would be just right to light them," the doctor said.

She crossed the room to the fireplace and lighted the six candles in the silver candelabra on either side of the white mantelpiece. Then she came back to her chair by the bed.

"Yesterday, just before you got here at noon, and she had begun to wander about a bit in her mind, she talked about the Sunda Straits. I remember she used to talk about them a lot years ago, but I can't seem to recollect now just where those Straits are."

"They used to be somewhere around the Indian Ocean," the doctor said, "in the neighborhood of Java. Somebody's always changing the old names for places, but that's where they were. I've read about them. They were a dangerous piece of water, full of shoals and reefs, with risky currents and sudden squalls. You had to weather them before you sailed up the China coast."

(35)

"Do you know about the Kerguelens, too?"

"No, I can't say I ever heard of them."

"They're islands," Lucy said, not a little proud of herself in the shadowy room, which pulsated with the racket outside as the tide flooded higher and the wind kept rising. "Or, really, they're great rocks jutting out of a waste of stormy water. When you once rounded the Cape of Good Hope on your way to Australia and when you once saw those rocky islands, you knew that with fair winds you could count on making Sydney Harbor in two weeks more."

"There was a lot of sailing to Australia right from this very coast after the clipper-ship days," the doctor said, "when the down-easters took over."

They did not speak again for some more minutes. When the hubbub outside ceased even for a few seconds, Lucy could hear the clock in the sitting-room ticking them away with the swings of its heavy pendulum. The candles flickered and wavered in the gusts which shook the window-frames, smote the walls, and now and again whirled down the chimney. The melting wax was running down their sides and making them untidy; but she took comfort in their being lighted.

The doctor got up from his chair and bent over the bed.

"She's just going, Lucy," he said.

Lucy got up, too, and stood beside him. He put his arm about her shoulders.

"She's smiling," Lucy whispered, tears blurring her spectacles. She took them off. "Just see her smile. Maybe she's seeing something she used to love. She's told me often about the rainbows they sometimes saw at sea with all the colors clear and going deep down into the water. There's never any such rainbows over just land, she said. Maybe she's seeing one of them now."

"Maybe," the doctor said.

6

Before she left the sitting-room at quarter past eleven, Lucy gave a slight twitch to the white fichu on Sarah Holt's black dress so that it would not look quite so severe. Even in her old age Sarah had always worn her clothes with a graceful, almost a jaunty air; and Lucy wanted the neighbors to remember her just as she had been. Then, when she had counted the chairs once more and straightened the window-shades, she went into the kitchen to get the basket of lunch she had brought over for Thaddeus. There were red geraniums blooming on the wide white window-sills. She broke off a blossom and stuck it in the cover of the lunch-basket.

Thaddeus was still pacing to and fro along the beach, which the tide was slowly flooding. He was in his dark suit which he had put on early that morning. His shirt was clean and his necktie a new one and carefully arranged. He was a tall man, well-built and handsome, like his father, his mother always said; but his large hands and his fingers were long, like her own. He was still twisting them nervously behind his back as he walked; but when he saw Lucy coming across the grass, he put them in his pockets.

"Your mother looks real nice, Thaddeus," Lucy said, as she handed him the basket. "And it's a pretty day after all that fog and wind."

"Yes," he said. "Thanks for all you've done, Lucy."

"Joel and Sam and the others will be over early to give

you a hand at the scow. They've got it all planned out, Thaddeus. You don't have to worry a mite. And the pails are all filled with water on the back stoop for the children's flowers."

"Thanks," he said again.

"I'll be going on now. The boats are getting close in, and Hannah will want to be fixing Ben's dinner for him. There's a big flask of hot coffee in your basket, and boiled eggs, and beef sandwiches. Try to eat it all, won't you?"

"Yes, I will."

She turned away to cross the grassy slope above the beach toward the gravel path which led through the field and rough pasture to the village road.

"Thanks for the posy, too, Lucy," he said.

7

The boats were close in now. Once she had traversed the path, noticing as she walked how bright the goldenrod and asters still were on either side and how red the mountain-ash berries in the thickets, she stopped by the open pasture-bars above the village road to watch them make the cove, drop their anchors, and load their dories with the morning's haul. Today, with the funeral close at hand, they would not take these up the Tidal River to the Pound there, but instead leave them in their own lobster-cars. She thought as she watched them cutting the quiet water, for there was hardly a breath of wind stirring, how many thousands of times she had seen them, or boats like them coming in through all these years. Dim shapes piercing the summer fog, or caked with ice from a sudden onrush of November sleet, or clear and distinct in a brisk northwest wind, and always, whatever the weather, with a cloud of expectant gulls above and around them.

They usually made the cove in much the same order, Sam Parker's *Lucy and Joel* ahead. Sam fished off the ledges on the far side of Shag Island, his bright red buoys on the water there lending color to the ranks of dark spruces crowding the shoreline. Young Carlton Sawyer could have overhauled Sam if he had wanted to, even though his far more dangerous grounds lay seven miles out, not far from the great light, where the Whirlpool Rocks rose black and sinister and the channel waters raced, on a high rip tide. Carlton's new boat, the *Mary Blodgett*, was the envy of the cove and even of a considerable portion of the coast with her sharp, slender prow and her fresh white paint. She was paid for, too, though how Carlton had managed with a wife and two children and a couple of poor seasons, no one could quite make out, except Lucy, whose heart always quickened with pride and pleasure when she thought of the young Sawyers. But there was an active code among fishermen, at least concerning precedence, and Carlton often slowed down his engine or stilled it altogether when he saw that he was overhauling Sam.

Benjamin Stevens had already rounded the high point of Herring Head, for he fished to the Westward in the long stretch of relatively shallow water which ran back of the Head into Mackerel Bay and made its way up the coast into a score of backwaters and estuaries. Ben's boat, which wallowed a bit in the best of weather, came on within the black shadows of her heavy gunwales which lay close but clear in the light of this incredibly bright September noon. He had painted her gunwales to match his black-and-white-striped buoys at the time when he had changed her name from the *Cormorant* to the *River Jordan*. This change, which followed his conversion some years ago to the small congregation at the head of the Tidal River and his baptism in the bitter waters there, had resulted in a lot of good-natured chaffing from his neighbors. When he was not about, the comments upon his new way of life and thought were apt to be less merciful, though one and all attributed the renaming of his boat to his wife rather than to him.

Daniel Thurston was sick in his small red house, which stood precariously on some ledges in a clearing halfway down the two-mile length of Herring Head, and therefore could not bring up the rear of the small fishing-fleet as he usually did. He always stood, tall and stooped in the stern of his boat, which was now little more than an oversized rowboat with an outboard, just as he always stood to row his dory, pushing her toward his beach with long backward strokes of his oars. Since he had lost his larger boat to his harassed creditors of the Packing Company eastward, to whom he had earlier lost his weir, he now dropped his small line of traps with their orange buoys much further inshore than did anyone else, rarely venturing outside the tumbling, surfswept boulders which marked the tip of Herring Head. Lucy missed his customary figure standing up there, black and bent, in his unseaworthy craft with Rover, his dog, sitting upon his haunches in the bow. When things were over this afternoon, she and Joel must go up to see how he fared.

Nora and Seth Blodgett took his place today, coming down from the Tidal River which flowed deep and strong between Shag Island and the cove. They ran a good piece

outward before they made the turn against the incoming
tide to swing in behind the other boats. Nora held the tiller
in the stern, an old felt hat of Seth's crammed down over
her head. Seth sat crouched over the engine as he had done
for far too long, Lucy thought. The Blodgetts fished up the
river now, after Nora had finally come to telling Seth that
it was there or nowhere so far as she was concerned. Yet
with lobsters the unpredictable creatures that they were,
they often made hauls fair enough even in inside waters.
And with the Pound nearer at hand, they could drop their
catch every day on the way down-river and home.

Now Lucy could hear echoing through the thin, bright
air the sounds of all those old, familiar, stubborn acts and
labors of men who harvest the sea: the clump and the
crunch of rubber hip-boots; the banging and shifting about
of gear in the holds and on the narrow decks, gaffs and
crates, bait-tubs, traps brought in for mending; the clank
of anchors and anchor chains; the swash of skiffs and dories
brought alongside; then the running out of loosened ropes,
the grinding of oarlocks, and the skillful, rhythmic dip of
oars.

8

She did not cease to wonder at the day as she now walked
quickly down the rough, uneven, narrow road toward the
few houses and the store. Here it was, contrary to every pre-
diction and to most experience, set between a week of
clinging fog and the certainty of the approaching equinox,
like some rare jewel discovered suddenly in a dark mine.
She scanned the southern horizon, far beyond the Whirlpool Rocks and the great light, for a familiar low white bank

of distant haze, but there was none. Sky and sea met in a radiant, cloudless line, beyond which as the crow flies, so Sam Parker said, the nearest land due eastward was the coast of Spain. Northward, behind the harbor, the sharp rise of the land met an equally tranquil, cloudless sky, as did also the wooded heights of the headland to the west, and on the east, beyond the wide opening to the Tidal River, the long, lonely, high shores of Shag Island. The whine and whirring of locusts and crickets filled the air. They took the place of the white-throats, silent now for a month and more; but Lucy was glad to see some migrating thrushes scurrying about in a clump of witch hazel before she reached the first of the few houses of the cove.

These were all small and of indifferent design. One was made from a schoolhouse, which years ago had marked a district sufficiently populated to warrant a school of its own. Another, so people said, had once housed a ship-chandler. A third, as rumor had it, had been a church; and in its case belief could take the place of hearsay since the lower half of what had clearly been a belfry tower sat astride its roof-peak. The three or four others, not so clearly designated, differed but slightly except in size from the fish-houses which stood across the road and in which the fishermen made their new traps during the winter months. Not one among them all had been rebuilt or built with any thought of composition, form, or contour. Each was merely a shelter for tired men when they came in from fishing and for the women who took care of them. Yet all were scrupulously clean without and within; and their small garden plots, set about with beach stones and brilliant with hardy flowers, petunias and marigolds, nasturtiums and zinnias, relieved their brown, or dark green, or weathered clapboards and shingles, their single narrow chimneys, and their ill-proportioned roofs.

Here and there above this meager settlement, in scant clearings on the northern hillslope, were half a dozen smaller houses, mere cabins or shacks, roughly covered with asbestos tiles, or heavy tarred paper, or ill-matched pieces of tin sheeting. These had been erected in off hours before the boats were again put into the harbor water in the spring, when the new traps were framed and fitted and there was plenty of time for such puttering. They would be occupied now and again during the summer months by men from coast towns who came for Saturday and Sunday to cast for pollock, a spirited enough fish to catch though unfit for anything but drying and not too much favored even for that. Still, both men and pollock had their uses, especially in a season when both lobster and herring were scarce, for a fisherman with a good boat could gather in twenty dollars for a day's outing. And sometimes when the erratic, whimsical herring did rush the weirs, and larger boats, intent on buying or scaling, filled the cove, there were quite likely to be some strangers among their crews, who, sick to death of filthy, stinking cabin bunks, were willing to pay well for the indulgence of stretching out their stiff legs for a few snatched hours in decent beds.

Directly below these makeshift shacks and across the narrow road from the village homes lay the beach, backed by the several fish-houses and by the piles of lobster traps and buoys between them. The tides ran high and strong here from the open sea which pounded on the jagged rocks of the Head and against the far side of Shag Island. They had long since hollowed out the shoreline of the cove into a deep, concave basin, nearly half a mile in width. Here, before the shingle and the shale began, the hummocks of hard soil in front of the fish-houses put forth intricate tangles of weeds and flowers, nettles, sea goldenrod, tansy, asters, purple thoroughwort, and the paler, more frail purple of sea lavender. The shale and shingle sloped sharply to the kelp-entangled, many-colored stones of the beach gravel, which in its turn gave way to the sand and the mud of the flats, where at low water, in the pools left by the tide, gulls struck after mussels or small crabs, herons stood, and sand-pipers cast their quick reflection as they wheeled in twittering half-circles. The beach was a disheveled place, given over to hard labor as the skiffs, dories, and scows, now floating free, but twice in the twenty-four hours imbedded or atilt in mud and ooze, gave proof; and yet it had its charm, especially on such a miraculous day as this.

To Lucy Norton, walking past the Stevens' house, which was empty since Hannah Stevens was keeping store for her and Benjamin was just now trudging up the beach with his gear, past the Blodgetts', empty also, the houses of the cove seemed neither undesirable nor ugly. Little or nothing in her experience had afforded her any standards of comparison as to line, or form, or color. They were simply houses where lived the people whom she had known for most of her life and whom she inconstantly dreaded, pitied, admired, worried over, but always loved. She was, instead, thinking of how gay the flowers looked in their garden plots and what a nice idea it was for Carlton Sawyer to repaint his buoys that bright blue with yellow handling-sticks. With the Stevens black and white and the Blodgett dull green, that splash of vivid blue against his fish-house would add life to things just as did Sam Parker's red a bit farther on.

She was still wondering, too, about the day, trying to discover in her mind how it was that sunny and windless hours such as these invariably restored the cove once more to its owners, to those who wrested their livelihood from it and from its outside waters. In the late fall and winter and again in a halting, capricious spring, there were days and even weeks when it was snatched away, belonged to no living being, but only to the winds, the cold, the fog, and the tides. These mighty, cruel forces took it back to themselves, scornful of their brief loans. So that whenever, during those long seasons of loss, you tried to recapture it, perhaps through seeing the tiny black shadows of the swallows, returning in the white mist to their homes, or perhaps by watching on a clear morning a curl of blue smoke from a smelt-tent on the frozen Tidal River, you could feel even then only an instant of ownership in the enveloping and sinister sense of having nothing at all, of being no one in a vast gulf of emptiness.

9

Mary Sawyer was standing across the road from her house, among the blue lobster buoys. She was painting them quickly and, when one was done, propping it against the sagging porch of the fish-house, against some fish-flakes, or convenient rocks. She was a tall girl in her late twenties with wide blue eyes and a generous mouth. She wore a faded pair of denim overalls and an old red mackinaw of her husband's, and both were flecked with blobs of paint.

"I'm making an awful mess, Lucy," she said, "but it's fun on a day like this."

"They're pretty," Lucy said. "I always did like that shade of blue. With so much on my mind yesterday, I never once noticed them."

"They weren't here yesterday. I began them just this morning. They're a surprise for Carlton. He's likely to be furious, but I got sick of these faded old yellow things."

She balanced her brush and her can of paint on the top of an uneven post and pushed back her short, fair hair with a sweep of the inner side of her arm.

"How does old Mrs. Holt look, Lucy?"

"Lovely," Lucy said. "She looks just lovely."

"What's she got on?"

"Her black dress with that white fichu around her neck. I washed and bleached it yesterday. I got it real white, too. It looks all fresh and nice, just as good as new."

"I'm glad," Mary said. "I'm glad of the weather, too. Carlton kept saying all through his supper last night that he couldn't see managing things just as she wanted them in all that fog. When we got up this morning, he couldn't believe his eyes, what with the radio report and all."

"Those weather men don't know a thing about us way down here," Lucy said. "Do you know where the children are? They ought to be back with their flowers by now."

"They went out on the Head nearly two hours ago. The little Randall girl said there were some of those red lilies in a swamp beyond Dan Thurston's place. I never knew those lilies blossomed as late as this, but she said she'd seen them, and she was set on getting them. She's a great one for wandering about all by herself. I didn't just take to my two going all that long way with her, but I couldn't say No with things being what they are, and they all that excited, and she such a forlorn little thing. I don't suppose the West girl minded about her little boy, but Hannah was terribly upset. She didn't want her grandchildren to go at all."

"It won't hurt them a mite," Lucy said. "I put some crocks and pails on the back stoop for the flowers. I told Thaddeus."

"Is Thaddeus all right?"

"He's fine. He's just as fine as he can be."

"That's good," Mary Sawyer said.

"I'll be going on now. I thought an hour ago that I'd never get my chores done in time. But the day and all has sort of quieted me down, and I don't feel flustered any more."

"I know. That's why I painted the buoys. I just let things in the house go and came out early in the sun. We can't hope for many more days like this one."

"There's always October," Lucy said. "There's usually a fine, still spell of weather in October."

Mary Sawyer reached for her paintbrush.

"October's awful near the winter," she said quietly.

10

When Lucy reached the store, which stood at the end of the line of houses and a few rods beyond the last of them, she found Hannah Stevens waiting for her on the small porch in front. Hannah was a thin, spare woman who looked underfed and frail. Joel Norton sometimes said in a burst of imagination quite foreign to him that he often wondered where Hannah found room enough to keep all of herself.

Lucy wondered the same thing today, for Hannah was clearly in an excited state. For one thing, her husband's dinner would be later than he liked it; for another, she was upset over the whereabouts of her grandchildren, who were spending an uneasy fortnight with her. But, resolutely thrusting these anxieties, with others more distressing, into the background, she first discharged her obligations to Lucy.

"There's been no trade to speak of," she said. "Some clamdiggers going through to Mackerel Bay got three bottles of strawberry soda. That's thirty cents. The West girl got a loaf of bread and two packages of cigarettes. That's fortynine. She charged it. All the money's on the counter with a slip of paper setting it all down. And that Randall child bought ten cents' worth of licorice for herself and the children before they started traipsing all over the Head. She paid for them, too. I'll have to say that."

"I hope you gave her twelve sticks for her dime. We always give six cents' worth for every nickel. And there's six children, you know."

"I didn't. I gave her just ten for her ten cents."

"I'm sorry," Lucy said. "She's a nice little girl."

Hannah looked annoyed.

"Well, it's a free country," she said, "or at least they say it is, and everybody has a right to their opinions. But my grandchildren have been raised up nice, and I'm not anxious to have them lose all their manners when they come down to stay with me. If it hadn't been for the funeral, I'd have put my foot straight down."

"Nobody's putting their foot down today," Lucy said. "This funeral belongs to all of us, even to the children."

Hannah said nothing for a moment, and in that moment Lucy was reminded of a magic lantern she had once seen as a child, how it first threw a black square on a white screen, and then, when the person who worked it slid some sort of shutter, filled the square suddenly with a bright picture.

"I made a cake this morning," Hannah said, "just to have it handy in case you need something extra for the men when they get back from Shag Island. I knew you wouldn't have a speck of time for baking."

"That was real thoughtful of you, Hannah," Lucy said.

She went up the three steps of the porch toward the door. The bright picture on the screen went out, and the black square came again.

"What about Thaddeus?"

"He's fine. He couldn't be better than he's been for two whole days."

"I should hope so," Hannah said.

She gathered up her ample wicker basket with its balls of heavy twine for bait-bags, with her mending, and her fancy-work, and started down the steps.

"Do you suppose Nan Holt will have the nerve to come, being it's Saturday and no school?"

"I wouldn't know about that," Lucy said.

She went into the store and behind the counter on the left to her familiar straight-backed chair. Habit strong within her, she put the forty cents, the quarter, the dime and the nickel on the counter, into the cash drawer which she locked. Then she hid the key behind some cans of soup on the shelf.

She sat down in her chair. Through the big front windows she could see the tide making up over the sand toward the broad line of beach pebbles and the shale above them. The fishing-boats floated at anchor; the dories and skiffs were inshore; the black posts of the herring weirs with their drooping brown nets were sharp and clear in outline. Down the steep hill which climbed toward the great headland she saw the children coming with their flowers. Some vivid spots of red in the armfuls which they carried proved that they had found the late lilies in the swamp.

On the high shelf behind the store the clock, which had stood in her mother's kitchen when she was a child, told her that it was long past noon. She had more things to do than she could ever manage to do before it should be time to go to Sarah Holt's funeral; but for the moment, sitting there, she could not for the life of her remember what the first of the many things was.

PART
TWO

The Neighbors

❖ ────────────────────────────────

Samuel Parker

◆ ————————————————————

SAMUEL PARKER got up early, even for him, on the day of old Mrs. Holt's funeral. He needed to be well underway long before the receding tide began to rush full strength down the mouth of the Tidal River and make things edgy for him in the fog when he swung across to the northern tip of Shag Island. He usually rounded the southern head of the island, going three miles seaward to do so, for his traps were nearer the outer ledges there; but today, with other things than just hauling to manage and the longer course to take, he would do wisely to get things going well ahead of time.

When he opened the front door of his house, which stood between Drusilla West's on the left and the store on the right, and stepped out to note the weather as he always did, he thought he had been transported to another planet while he slept. He expected the same clinging fog which for a week had driven them all crazy, making them crawl homeward three hours late, demanding compasses and even soundings; he expected the same southwest wind, rising all day long and shifting suddenly all over the place. There was no wind. It had died somewhere offshore in the night. Nor was there so much as a hint of fog. The air was clear and dry. Some pale stars were pricking the sky. He knew as he stared into the darkness that, once the dawn had broken, the sea would be calm to the far horizon, and, when he had made his fishing-grounds, the towering shaft of the lighthouse beyond the Whirlpool Rocks would be seemingly so close that he could almost lay his hands upon it. Wonders would never cease on this coast, he said, as he went indoors to light his oil-stove and to get the water hot for his shaving and his coffee.

He was a deliberate, careful man, and his years alone had only strengthened his slow, methodical, tidy ways. The three rooms of his house were clean and orderly: his small bedroom on the right of the front door; his sitting-room opposite, with its air-tight stove and neat wood-box, his few books, and the table where he sat often in the evening over a solitary word-game which he enjoyed; his long kitchen at the back. The kitchen was divided, at least in his mind, into two distinct parts: one where he cooked, ate his meals, and washed his dishes; the other given over to a kind of workshop with a pine bench in the center, cans of paint on some shelves, and tools hanging upon the walls. Here in the winter, when his boat was taken up and canvased and his new traps were framed and ready, he whittled out small lobster buoys, which he painted in bright colors, even striping them carefully and boring the ends with tiny holes; made little traps, clam-baskets, skiffs and dories; or sometimes a fair-sized sloop or a schooner, complete with rigging and sails. He had found a ready market for these notions, both from occasional visitors, who in the summer bumped and swayed along the bad road in search of a picnicking place, and in two or three shops in the coast towns, to which he carried them in Joel Norton's truck at infrequent periods during the winter. The Nortons also displayed them for him in the store. He never allowed his fishing paraphernalia to clutter up his workshop or his kitchen proper. This, his hip-boots, his overalls, mackinaws, and oilskins, his lanterns, and his odds and ends of gear and tackle, he kept in a lean-to which he had built on his back porch.

He had the habit of talking aloud to himself as he went about his house or when he was alone in his boat. Since this indulgence troubled no one and afforded him company and pleasure, he did not see any reason for curbing it. Used as he had become to the sound of his own voice, he often even read aloud to himself, liking the rise and fall of words and sentences. It was doubtless these pastimes, together with his word-game, which had given him a fluency and correctness of speech not common to men of his calling and background.

"If I were a superstitious sort of fellow," he now said as he put two eggs to boil in a saucepan, "I'd be inclined to think that this day had been created especially for old Mrs. Holt."

He ate his breakfast slowly, the two eggs, some warmed-over biscuits, which Lucy Norton had made for his supper the night before when the fog had brought them in late, and plenty of steaming coffee sweetened with thick condensed milk. He washed his dishes before he went outside to light his lantern, though the dawn was beginning to break, get into his fishing clothes, and start shoreward. His

dory was floating free, and in ten minutes he was clamber-
ing aboard his boat, slipping her from her mooring, and
easing her eastward, as noiselessly as he could, through the
Tidal River channel toward the shelving ledges which
marked the northern shore of Shag Island. Once he had
skirted the ledges and was well in open water on the far
side of the island, he hung his lantern in his shipshape cabin,
lit his pipe, and took up his position behind his steering-
wheel.

The sea was incredibly still. Usually, after a blow of two
or three days, it did not quiet down for hours on end,
especially here in deeper water with the open Atlantic just
off his port bow.

"Again, as I said before, if I were a superstitious man,"
he confided to his pipe, his purring engine, and the dark
island spruces.

The eastern shores of Shag Island for fully half its three-
mile length were high and heavily wooded. Whatever open
pastures and rugged fields there might once have been, like
those still discernible on its western slopes, they had yielded
themselves long since to the steady, relentless conquest of
the trees. Firs and spruces had marched tenaciously downward from the crown of the island, which they now ob-
scured, to the boulders and the ledges at the water's edge,
and even in these they had somehow found a footing. They
formed a wall, or palisade, of almost impenetrable darkness.
Only here and there where, for want of room and sunlight,
one had died, did the ranks of massed green give way to
rust color or to the silver gray of some moss-hung skeleton.

The *Lucy and Joel* idled through the water some rods
offshore. Her skipper, now that he had made the tide, was

in no hurry. The sun was not yet up, but the eastern horizon was a pale, clear yellow, ready for its first long paths of light across the sea. These would send the shadows of his sparse rigging far toward the shore, irradiate the dripping spruces with millions of prismatic crystals, and bring into bold relief the first of his red lobster buoys but half a mile away.

2

Before he had come to the cove settlement nearly twenty years ago, Sam Parker had earned his living in a variety of ways, none of which precisely suited him. He had worked in Maine shipyards on vessels now destined quite as likely for foreign as for home registry and ownership; helped to man a fishing schooner bound for the Banks; learned the oily methods of packing herring in a Passamaquoddy factory; and trundled a freight barrow for the Eastern Steamship Company, which in his young manhood was still sending the *Belfast* and the *Camden* back and forth between Penobscot harbors and the Atlantic Wharf in Boston. He had shifted from one to another of these ways of life, never knowing exactly why he felt out of place and restless, but rather suspecting it was because he was too close to too many people.

Like thousands of other young men of the coast and islands he had enlisted in the Navy in 1917, only to be sent to the Great Lakes where even the gulls seemed lost and where the inland water had no smell except of tar and oil. His very uniform looked misplaced and alien under those distant, glaring skies, among the sand dunes and the flat, empty beaches; and after the ravages of influenza in the autumn of 1918, when he had seen far too many homesick boys die after far too few hours of fever and dysentery, he had shed it willingly enough for good and all, though such a decision was contrary to his earlier dreams.

When he thought it over, as he sometimes did on quiet mornings which allowed his mind to wander pleasantly, he knew that he had chosen to cast his lot on and off this weatherbeaten point of land largely because of Lucy and Joel Norton. They had all been young together on an island farther westward, although both Lucy Vinal, as she was then, and Joel were five years ahead of him in age. As a shy, awkward boy he had loved Lucy, and the love still endured if after a different, less upsetting fashion. And when, through the death of his parents and the departure of his only sister to California, he was left with no family ties, he decided, after an inspection of various other fishing communities, to settle down in this remote cove where the nature of winds and tides demanded whatever wits a man possessed and where the problem of too many people was nonexistent.

He could never have imagined in those nights off the Banks when, crammed among other spent and filthy men, he tried to sleep, or stood his watch against the dim riding-lights of other fishing craft, against roving icebergs, or the gigantic bulk of an ocean liner, that he would one day know the security and comfort of his own house. He thought himself the most fortunate and contented of men. He made a decent living with his traps; and his whittling and carpentry could take up the slack of an especially poor season. He had never married, not because he cherished any romantic notions of a first and only love, but because he had always hesitated before the picture of his life with any of the few unattached women he had ever met. To do him justice, he had been even more hesitant before the thought of their life with him and his somewhat stubborn, solitary ways. And whenever he was distracted by those urgings and desires common to all, he could quite reasonably count on any one of a dozen happenings provided by this merciless, treacherous sea to take his mind off his deprivations and set him straight again.

3

The sun had just swung clear of the horizon when he reached the first of his red buoys and began hauling in his traps. He did this with the slow, steady rhythm of the long-practiced fisherman: balancing each against the narrow stern deck, emptying it of its squirming freight, throwing back those which failed to meet the requirements of the gunwale measuring-rod, rebaiting, casting each trap downward with a plop and swirl of water. The morning's haul was better than average as though the lobsters, pestered by deep undercurrents, had gathered in restless, writhing hosts near the island ledges. When he had plugged a couple of hundred menacing claws with bits of whittled spruce, he dumped his sprawling cargo into his bag-net and floated it alongside. Then he straightened his tired back and shoulders by slow jerks and swings of his body, rekindled his pipe, hoisted himself to his foredeck and, lying against the deckhouse in the early sunlight, carefully scanned the island shore.

Here the spruces were less thick and stubborn than on the northern half of the island. He could discern spaces of light among them and across this lower land catch a glimpse here and there of the sea on the other side. He thought he remembered from an occasional November hunt after deer that a swamp lay beyond these trees and that a rough path rose from it to the western shore, where there had once been, many years ago, a thriving settlement, ship-yards and docks sloping to the deep waters of an outer cove. He knew that place well, and especially since the late afternoon of yesterday when he and Joel Norton and Carlton Sawyer had crossed over with their spades to dig old Mrs. Holt's grave in a long-deserted family burying-ground there. They had done their work in a clinging fog, finishing it by the uncertain light of lanterns propped against or resting upon the few tombstones, coming home wet to the skin, and sad besides. He now studied the geography of the island only because, on account of his secret approach to his traps from the north instead of from the south, his access to the place must necessarily be a different one.

"I'm quite sure of that swamp, however," he said now to some light, vagrant winds. "And if I'm not mistaken, the path peters out after about half a mile, just in front of those old cellar holes."

When he had finished two pipes, he went below and took off his hip-boots, changing into a pair of worn sneakers. From the deckhouse wall he took down his axe, testing its blade as he did so. Then he hauled in his anchor, started his engine, and, mindful of the receding tide, made for deeper water. In a few minutes he was in his dory and rowing toward the red ledges, which already showed a helpful strip of beach at their base.

He had been quite right about the swamp. Though it was

now choked with alders, brake ferns, and the spiked leaves of iris, it was still a swamp. As he lunged through it, searching for better footholds, gulls, surprised at this invasion of their solitude, screamed above his head, and an osprey wheeled suddenly into the air from its untidy nest on the top of a dead spruce. He was drenched to the knees in black mold when he at last found the overgrown path, which wound upward through more masses of alder and huckleberry, bayberry and sheep-laurel. Before he set forth upon it, he looked back upon the swamp.

"All those irises in the spring must make a pretty sight," he said. "I must bring Lucy over some day to see them."

The sun told him that it was going on for eight o'clock when he had finally left the obscure and tortuous path and arrived at the cellar holes, which only the battering winds, from the more open sea, smiting the island summit and discouraging the trees, had saved from the complete obliteration of the houses once standing upon them. But he still had time for two hours' work before he should make his way back to his boat and start her up to round the southern end of the island and set forth for home.

A hundred yards or so below him, among the heavy undergrowth and almost buried from sight within it, rose the rusted iron fence which enclosed the burying-ground. He could see, as he drew closer to it, the evidences of their work the night before, the heap of brown, rocky earth with their spades standing upright within it. They had moored at some black posts, the surviving relics of piers, and, once ashore, made their way upward across slippery, sloping timbers, brown and green with ooze and slime, half buried in shale and gravel. It was difficult to believe that the great hulls of ships had once slid down them into the deep, full tide while eager Shag Island people watched and cheered. Had a cannon shot rung out, he wondered now, as a warning signal just as the blocks were knocked away and a brig, barque, or even clipper, destined for the utter-most parts of the earth and the sea, started downward, sliding faster and faster amid a cloud of dust and flying splinters?

"They always fired a cannon shot at all the launchings," he said, as he neared the iron fence. "It's hard to think of a cannon over here on Shag Island, and I can't recall now that she ever told me about it; but I'm sure there must have been one."

He began at once to clear away the matted growth within the burial enclosure, working with his hands as well as with his axe and throwing armfuls of tall, brown grass, tough bushes, and small trees beyond the iron palings. There were five tombstones. The largest, which bore the date 1852 in raised granite figures, was still upright. The other four were aslant or fallen in the dead, neglected grass. He wanted to raise at least one of these up and prop it with big rocks; but he could not do this with just two hands. He must be content with clearing them all and with tidying up the ground itself.

When he had accomplished this to his scrupulous satis-faction, he sat down on the gray granite coping which surrounded the graveyard and in which the iron palings had once been carefully placed. He examined the base of several of these and saw how firmly they still held.

"The man who did that job," he said in quiet admiration, "did a good one."

The sun was now higher, swinging a bit southward in its narrowing autumn course. The air, motionless among the surrounding trees, vibrated with the hum of a myriad unseen insects. A flock of brown curlews, rising from some hidden feeding-ground, swept overhead, uttering their sharp cries from their curving beaks. They were staying longer this year than usual. Perhaps lonely, outlying islands tempted them to delay their northern return. He was thinking as he smoked his pipe.

"I hadn't reckoned until now," he said slowly, watching the frail blue spirals float upward, "that I was doing this for anyone except her. I'd thought of it as a sort of token for all she's given me. But I see that I'm doing it for all who once lived here, for the fellow who set these iron pickets and for the men who dug those cellar holes and built the ships. And there's more to it than even that. I'm doing it to stir up the past and all the things it meant, once upon a time, to places like this."

After a few minutes he began to wield his axe almost savagely among the growths outside the burying-ground. Small spruces, pines, and firs, tamaracks and birches, cut close to the earth, fell before his sure, sharp strokes. He saved a clump of swamp maples, which had already turned scarlet, just as he had spared a single blue harebell which was in blossom against one of the fallen stones. Within an hour he had cleared a space of nearly twelve feet on every side and, in addition, had cut a rough path through to the landing where the timbers were. He should have brought Joel's scythe, he thought regretfully, as the sweat darkened his blue shirt; but he could hardly have managed both scythe and axe in the swamp and on the overgrown path upward.

When he had gathered up his heavy cuttings and made a huge stack of them as far away as possible, he walked slowly around his new clearing. Now the utter desolation of the place had vanished. Remote and solitary though it was, it was no longer completely abandoned and unkempt. And when the spring came and he brought Lucy over to see the irises, they would cut the new wild grasses in the burying-ground, burn his piles of brush, and perhaps even bring some mortar with which to raise and make secure the fallen stones.

There was just one more thing to be done before he made his way back to his boat. This he had determined upon during the night when, sleepless over the desolation of the island graveyard, he had formed his plans for the morning. He stood in his clearing and scrutinized the still encroaching thickets and underbrush. Up the uneven slope, just to the left of the cellar holes, some mountain-ash trees grew, laden with their broad, flat clusters of red berries. He was elated as he strode toward them. They were precisely what he wanted. Old Mrs. Holt had especially favored them. She always called them *rowans*, the Scotch word for them, she said.

He felled the huge clump at its base, arranged the stems carefully in line, and dragged them down the slope. He covered the unsightly mound of rocky earth with them and lined the edges of the open grave. They looked nice there, he thought, in this new sunshine for which he had afforded space. They eased his anxiety even although there would be no others at the burial except the men who would maneuver the narrow scow through the ebbing tide at the mouth of the Tidal River, tow it down the cove toward the black posts and the rotting timbers, and somehow moor it there.

"And all that is bound to be a tricky business," he said, "so close on the edge of darkness."

Then he put on his corduroy jacket, took up his axe, and started toward the opening of the path which led downward into the swamp.

Lucy and Joel Norton

◇ ———————————————

LUCY VINAL and Joel Norton grew up together on a sizable island in the western waters of Penobscot Bay. The island was, at the time of their birth and childhood, largely given over to the quarrying of granite, with fishing, in comparison, a minor industry. They had been born far too late to see, or even to be told much about, those fishing-fleets which, throughout the first half of the nineteenth century and earlier, had sent out from most Maine harbors, whether mainland or island, their broad-hulled, sharp-sterned pinkies and their tubby, barrel-sided schooners bound twice a year for the teeming seas off Labrador and Newfoundland. They had never known those prosperous merchant-shipowners of the larger coast towns who in former years relied upon the fishermen to supply countless quintals of salt cod for their cargoes to the West Indies and even to southern European ports. In their childhood the few craft which still sailed to the Banks or to the nearer waters of the Bay of Fundy were manned by that Maine composite of mariner

(68)

and farmer, men who in the spring, between sowing and late summer harvest, and sometimes again between October and the setting-in of the long winter, ventured forth in their home-rigged boats to increase their livelihood by selling their still generous catches to Rockland, Portland, and Boston markets. Their own fathers were of this breed and mixed calling.

While they were studying their dog-eared books in the indifferent island academy, the granite quarries were likewise beginning their swift decline. Since masons in the cities were pouring concrete instead of hewing granite blocks into shape for churches, fine houses, and public buildings of every sort, there was less demand than there had been, even a brief generation ago, for the great gray slabs of native rock which nature had lavishly provided along so many island shores.

In place of both fishing and quarrying a new industry had arisen along the Maine coast from the wide, white sands of Old Orchard to the deep, landlocked harbors of Frenchman's Bay, the sure and safe, if less adventurous industry of catering to the needs and desires of summer residents and sojourners. Maine families, whose surnames a hundred or even fifty years earlier had been known far and wide in the Indian Ocean and along the China coast, adapted themselves to this change in events with relief, if not with eagerness. Summer people with capital meant a new spur to local business. If worse came to worst, great houses, built with the proceeds of foreign trade, could, conceivably, be sold. Their sons and daughters would find yachts to man, horses to drive, lawns and gardens to nurture, tables to provide for and serve, and city children to care for. As soon as Lucy Vinal and Joel Norton had graduated from the academy, they found waiting for them any one of a dozen jobs, none of which they really wanted.

Yet even if Joel Norton had been born fifty years before, in the days when a score of lively New England seaports from Passamaquoddy to Narragansett were far better known in foreign harbors than were the greatest of American cities, he would hardly have added distinction to any of them. Like most sons of families long devoted to fishing, he was not of the stuff of which deepwater sailors and future shipmasters were made, belonging rather by background and tradition to that lesser and humbler form of marine enterprise. He was framed for habit, not initiative, for quiet independence rather than for the ruthless discipline of a forecastle or a quarterdeck. As his father and grandfather had done, he preferred stable to precarious ways. The steadying knowledge that home awaited him after a brief and relatively safe venture into tossing waters which, on the share system, might help to support him in decency if not in plenty, would have seemed more welcome to him than any dream of a first officer's berth and, eventually, a captaincy on an East Indian or an Australian merchantman. In any time or place he would never have been fired with vaulting ambitions.

Now that fishing with its initial outlay of provisions and gear had become more costly and, like granite quarrying, was distinctly on the wane, he turned with common sense, if reluctantly, to less attractive livelihood near at hand. At nineteen he became the skipper of a pleasure craft owned by New York people, who had built a summer home on the nearby mainland and who thought themselves lucky to have found Joel as, indeed, they were. They fitted him out with a blue coat with brass buttons and a jaunty cap of crisp white duck.

Joel was neither in temperament nor appearance at all jaunty. He was short and thickset with slow, cautious ways. He had reddish curly hair and round blue eyes which widened when he felt ill at ease as he often did. He was an excellent navigator, having known the coast and islands from boyhood; and he kept the small, spruce, yacht-rigged boat in the best of shape. During the three summers when he piloted her on short cruises or carried cargoes of young people on deep-sea fishing trips, he gratefully and haltingly confided his misgivings to Lucy Vinal, who served the same household as a quick and competent waitress.

Their employers were the kindest of people, they both agreed, when in their free hours they took a picnic lunch to some secluded cove or on a rare day off returned to the island in an equally spruce motorboat; and, so far as they knew how, they supplemented their generous wages with friendliness and consideration. But they were alien to the settled ways of coast and islands, whose people, long accustomed to sturdy self-sufficiency, found it difficult to share their native grounds with the best-intentioned of invaders, particularly with those representing equally alien social and financial circumstances.

"I don't quite know how to put it," Joel said to Lucy, fumbling with his smart cap and feeling his hands grow wet and his neck red and hot, "but they'll never belong here with all their trying. I feel sort of sorry for them,

(71)

though I'm sure they wouldn't thank me for it. But, come to think about it, I feel sorrier for us and our own old ways."

Lucy at these moments felt sorrier for Joel than for the summer people or even for the old ways. Whenever she saw him grow red and confused, searching for the words that lay in a tangled mass inside him and refused to come out of his mouth, she longed almost fiercely to protect him against all that made him taciturn and awkward. Just as at school she had helped him with his lessons, over which he was often discouraged and always inarticulate, so she wanted now to lend him confidence and at least a measure of ease. She was not overly contented with her own summer job; but she was by nature quick and responsive, able almost at once to extricate herself from momentary embarrassments, likable, outgoing, friendly, given to laughing over her stupidities rather than to brooding upon them. And during the three years, between the autumn and the late spring, there was the island school where she taught the children of her neighbors, while Joel stayed in one small wing of the big house on the mainland, taking care of things in such a worried, persistent fashion that he was always fearful lest something go wrong whenever he infrequently crossed the thoroughfare in his boat to see her.

Matters reached a climax in August of the third year of their summer work together. The New York family liked Lucy so much, her adaptable ways, her superlative service, and her cheerful good nature, that they asked her to return with them in the autumn. The work would be no harder than in the summer, they said, in fact easier with the added conveniences of a city house, and it would besides give her the chance to see something different from what she had thus far known. Lucy confided this news to Joel on an afternoon when they had gone in the motorboat to fish for cunners off some neighboring ledges.

When Joel heard it, he was seized by panic. It was as though his one anchor had given way and he was being swept toward perilous reefs and submerged, invisible rocks. This sickening sensation, the worst known to any mariner, so terrified him that somehow, in sheer desperation, he managed to dislodge enough words to beg Lucy to marry him and stay at home. And Lucy, swept by pity and as sure of Joel's goodness as of his need, did not keep him in his agony and terror more than half a minute. Nor had she ever regretted her decision for even a fraction of that brief period of time in the thirty years which had elapsed since that August afternoon.

2

They did not, however, stay at home on the island. Rumor reached them from the master of a seiner, chugging through eastern waters in search of herring, that a general store in an out-of-the-way fishing settlement had been vacated by its discouraged owner and was for sale at almost any price offered. The place, a hundred miles and more up the coast and in a region designed for fishing, alike from its physical features and its source of supply, appealed to them both. They were not made, they decided, for summer employment, ever on the increase, or, in Joel's case, for long winters of routine caretaking and of filling in spare hours with odds and ends of island labor.

On those vast stretches of deeply indented coastline east

of Frenchman's Bay there are still remote communities facing the open sea from isolated points or headlands, or up some tidal stream, or clustered about coves and backwaters, and quite off the track of the general summer invasion. Builders and owners of summer homes prefer safe harbors for their yachts and cruisers, nearer access to markets, to golf links and tennis courts, and to neighbors of their own sort. Only a few strays among them, intent on solitude at the price of comfort, have ever penetrated the most outlying portions of far-eastern Maine. Those distant regions belong now as they have belonged for two centuries to its native stock and, in particular, to those who trawl or seine, drag, or weir, or haul their treacherous, windswept waters.

Joel's circumspect, overcautious nature might well have hesitated before investing their slender capital in a dilapidated two-story building set in so sparsely settled a district, had not Lucy's fervent eagerness vanquished his prudence. Within an hour after their initial inspection of the empty store and its surroundings, which in themselves delighted her, she had imagined the building renovated to afford them a home as well as a place of business, freshly shingled and painted through their combined labors, and gay with flowers behind the wide front windows. While Joel carefully tested sills, roof, and a rickety flight of outside stairs, examined the outhouse, and discovered the whereabouts and condition of the well, she talked with a few curious residents.

There were many more people about, they said, than Lucy and Joel could see at first glance, they having come by water rather than by land. At least a dozen families lived along the road leading to the main highway, families of men who fished in those bays and backwaters which cut into the long, narrow point of land from east and west. The keepers of at least three light-stations used the store as the center of their trade, and there were people on outlying islands, too, who came in regularly for provisions. In a good season there were also any number of boats which made the cove their headquarters, however briefly. Hunters were a common sight in November; and more often than one would think in the summer months a launch or a yacht made harbor for the night and stocked up heavily on supplies. Had not the former storekeeper, they said, been an outlander from somewhere upstate and quite unused to fishermen and their ways, he might have made a decent living. But, there! it took a native coast-dweller and, better still, a fisherman as well to stick to things through thick and thin.

In the end—and there was not really an end, for Joel had no arguments against Lucy's fervor, or at least none which he could dredge up in time—they bought the store.

3

As a matter of fact, and very salient fact, indeed, Lucy Norton's inclination and even ardor toward shopkeeping rested upon a profound and tender understanding of her husband. This, like all wise women throughout all ages, she kept strictly to herself. She was, it is true, actuated by the pleasant thoughts of children clutching their pennies in anxious deliberation before jawbreakers, jelly beans, and black ropes of licorice, and of women consulting her concerning the relative merits of packaged and home-set yeasts. She liked the warm picture of tired fishermen smoking their pipes for an hour around the stove on winter evenings before they went home in the windy darkness. She liked to imagine well-stocked shelves, row upon row of brightly labeled cans, jars, packages, bags, and bottles. Yet these were all mere accessories to the fact, to the knowledge which lay deep and strong within her.

In partnership with her, Joel could run his store, overcome his diffidence, defeat his fears and hesitations, discover himself respected and honored, even as she respected and honored him. And if she watched herself as she knew she could and would, he might before too long come to recognize himself as sole owner and proprietor, with her merely as his helper and assistant. On the white signboard above the new front door, which she fancied as dark green against the fresh white clapboards, the matching green letters would say:

JOEL NORTON. GENERAL STORE.

Joel did not fulfill all Lucy's hopes and dreams for him, but he slowly realized enough of them so that, before too long a time had passed, he was more contented and comfortable in his mind than he had ever thought possible. It was not difficult to overcome shyness, or at least to be less aware of it, in the company of men naturally taciturn and withdrawn. The very fact that he was not required to say much loosened his tongue. Now and again he modestly ventured an opinion on something besides the weather or even proffered a bit of information during an evening. Perhaps on the despised mussel:

"I'm bound to think the time's not far off when we'll be payin' heed to mussels as to clams. I've been readin' how they're a prime favorite food in some of the countries of Europe."

Perhaps on the droppings of sea-birds:

"I'm inclined to think we're missin' the chance to get some first-rate manure for our gardens right off our own islands. It's feet deep in some of them crevices, and we could load a fair-sized scow in no time at all, workin' together and given a quiet tide. That's what our deep-water sailors did years back. They sent big ships down to some islands off Peru in South America just after bird droppin's. It don't seem sensible, but it's a fact. They called it *guano*, and they carried tons of it as far away as Europe. I understand it fetched big money, too."

When Lucy heard Joel's voice confidently saying these things and saw how respectfully his neighbors listened to him, her heart leaped within her. She would keep on reading the books which old Mrs. Holt in her house at the eastern end of the cove had urged upon her and continue to share bits of them with him over their bedtime tea and toast.

During their first ten years of storekeeping things often looked dark enough. Theirs was too small and out-of-the-way an establishment to allow them the wholesale markets with their lowered costs, even if they had had any access to these. Supplies must be purchased in the nearest towns, reached by sea until trucks became common even on bad roads or before it was possible to save the money to buy one. Profits were pitiably small at first; and always, from the beginning, credit must be cheerfully extended during the winter months following a poor fishing season. Still, the prophecies uttered at the time of their initial visit proved, on the whole, sound. The scattered families living on the long stretch of road must have the ordinary staples, plus potatoes and salt pork to serve with their omnipresent diet of fish. Hunters, bound for rough shelters on Herring Head, or Shag Island, or in the woods to the north, stocked up heavily with supplies. The light-station and island patronage, while never heavy, was by no means negligible. On days when the weirs were bursting with fish any number of boats filled the cove with hungry men. And when, finally, a tank was sunk for gasoline and the gas truck got through regularly, trade in fuel for boat-engines and in fat cans of motor oil made the future look bright indeed.

Joel fetched and carried the goods, at first going once or twice a week in his boat to the town most accessible by sea, taking with him the long list which they had made out together during hours of careful deliberation. Later on, when he had his own truck, he went daily at dawn to the nearest towns, and frequently, as their business grew, to the farther wholesale markets. Lucy did most of the selling. Joel was always disturbed by that. Whenever Hannah Stevens in her brisk way reeled off a sizable list of needs all in one breath, rice and raisins, baking-powder and sugar, pipe tobacco and Quaker oats, his neck grew red, his fingers holding the pencil trembled, and he became hopelessly confused with the adding up. At such times Lucy always remembered something she needed upstairs which only he could find for her, or supplies in the back storeroom which were far too heavy for her to lift.

4

Thirty years, Lucy often thought now, as she swept the store in the morning and got things started for the day.

Thirty is an awful lot of years, almost half one's life. Have we really been here thirty years?

She tried to remember what she and Joel had looked like when they had come, but that picture faded before what they looked like now, at fifty-three. Joel, more than a little stooped from sitting long hours in his truck and from carrying sacks and cartons; she, with graying hair and fine lines crisscrossing her face, unmistakable lines which she did not much like. Not that she regretted the years. She was merely now and again stunned at the discovery of them, piled up so irrevocably behind her. The two of them had gone away infrequently, to be sure, always flurried and eager over the idea of a week elsewhere, even at the cost of closing the store—to see the wide fields of the Aroostook or to savor the excitements of Bangor or Portland; but all things after a few days had taken on an alien confusion, and they had been relieved to return to their familiar ways.

The cove settlement had actually changed little in these years. Fishermen of the most stalwart breed among that calling are not by nature rovers. Having once selected a plot of ground upon which to live and an expanse of water within which to labor, they are likely to cling to both like tenacious barnacles upon rocks. Well aware of the vacillating, mercurial, unpredictable ways of herring and lobsters, they wait upon these, trusting wisely to the fickleness of inconstancy itself. When the cove fishermen heard of filled weirs not three miles distant or learned that lobsters were moving eastward or westward to enter traps not their own, they merely shrugged their shoulders over the irritating fluctuations of chance and awaited their certain, if delayed, turn of fortune.

Hannah and Benjamin Stevens, now past sixty, had not pulled up their anchors for over thirty years, nor had Nora and Seth Blodgett. They had been here before the Nortons had bought the store. Daniel Thurston, old and frail now, on his beach lying under the shadow of the headland, boasted that he had known the cove for half a century. The dwellers in the other houses had changed, sometimes by reason of sadnesses which Lucy tried to forget, but others had taken their places. Sam Parker, for going on twenty years now, seemed content and afforded Lucy and Joel quite inexpressible pleasure, hunting with Joel in November, going with him during the winter on his buying trips, to Lucy's unspoken relief, lending a willing hand at countless jobs. The young Sawyers had lived here since their marriage ten years ago, taking over the house of a fisherman who had drowned when his engine failed him in a gale off Herring Head. Drusilla West, known as Trudy, had been recently left, an unwelcome bit of flotsam, by her husband who had been brought up here as a boy. He was working now on a freighter on the Great Lakes, and he might, or might not, return for her. The Randalls with their one child had slunk into the village from God knew where two years ago, bought an acre of land from Daniel Thurston, and erected a house of sorts on the long hill leading to the headland. They were seemingly not dependent upon fishing as a livelihood.

5

On that September day, now so long ago, when they had finally come to settle in and while they were impatiently waiting for the tide in order to get shoreward as far as

possible with their laden boat, Lucy from her station forward among their piles of household gear had seen a tall
woman walking back and forth above the eastern beach
and now and again scanning the cove waters through a pair
of binoculars.

"That must be that old Mrs. Holt," she said to Joel.

Joel was too anxious about the unfamiliar harbor, the
scow he was towing, and whether he could presume upon
his new neighbors to help him unload, to give more than
a cursory glance toward the beach.

"She walks different from most coast people," Lucy said
to herself. "They called her old, but she doesn't look a bit
old to me."

Sarah Holt had never seemed old to Lucy Norton from
that day onward through all the years. Her kitchen and her
sitting-room had from the beginning been Lucy's magnet,
drawing her away from the store in every rare hour she
dared to snatch. There in the old house while they hooked
or crocheted, knitted or mended, or roughened and cut
their fingers on the endless fashioning of bait-bags, Lucy
learned of places she had never known existed, of people
who had been but a name, and of thoughts which, unaided,
would never have opened her mind. She began to read
books after a manner which her years at the island academy
or her school-teaching had never taught her.

"If it hadn't been for books," Sarah Holt said, "I'd have
been in the Tidal River long ago."

Lucy loved these occasional wild remarks. Once in a
while she tried one out on Joel, who looked alarmed.

She was transported into a new world in Sarah Holt's
house. Yet it was not alone the world of the past, a world

of towering, wind-filled sails which had once spelled adventure and triumph. It was far more often Lucy's world of the present, the immediate narrow world of a cramped community set down on a barren, sinister coast, the somber beauty of which only heightened its desolation. Sarah Holt was able, perhaps through the very magic of the past, in some mysterious way to transform that world, lend hope to it, endow it with realities, with values and new meanings. It still remained a world of dirty engines, hungry for gasoline and oil; of tough spruce withes bent for lobster traps; of costly nets too often empty of fish; of obstinate, durable, weary men fighting stolidly against all odds of wind and weather; of anxious, envious women studying mail-order catalogues; of children who early knew more about danger than dreams. But through Sarah Holt's tenacious hold on life, its unremitting labor became curiously dignified, its people raised to a stature of which they themselves were unaware.

The woman who could transfigure for Lucy Norton this meager sphere which she had so impulsively chosen as her own, untangle its puzzling intricacies, relieve its obscurity, lighten its frequent darkness, was herself the heart and center of that sphere. She was like the oaken keel, the backbone, of one of the ships she had known, chosen and hewn to hold its frame invincible against wind and waves. The past had given her knowledge and experience, to be sure, and these enlivened her thoughts and afforded her solace and pleasure; but above all else it had granted her that intuitive wisdom which is the greatest gift it has to bestow. In her long life she had known several differing worlds, and the changed face of each succeeding one had afforded her wonder, regret, curiosity, dread, and hope. She at once cursed and blessed them all.

It never occurred to her that the present inconsequential community in which she had lived much of her life, through force of circumstances, either shut her away from places outside it or was in its essence different from them; yet she knew the peculiar environment which determined its nature. She knew, therefore, those with whom she lived, her immediate household and her few neighbors. Like all people they were unfathomable composites of conflicting emotions and desires; and yet these were made a thousand-fold more conflicting by the very character of their situation and occupation. She was not distressed by the frequent discovery within them of motley, contrasting qualities, for she had learned that malice and charity, tenderness and cruelty, pettiness and nobility can exist almost at the same hour in most human hearts. She was, in addition, quite able, with regret, but with humor, to detect them likewise within her own.

On those days when the cove, islands, and headlands, and even the great light retreated suddenly into the fog, and everything tangible and visible seemed to return to the sea and the winds; or when a flock of geese against a gray autumn sky brought senseless dread, and the mad laugh of a loon impelled one to go away, back home, to the kitchens of summer hotels, anywhere at all, Lucy would recklessly leave Joel to mind the store, and start eastward along the village road. She took the same way when, after empty weeks, money was scarce, people grew restless and querulous, and murky rumor whispered of trap-lines being revengefully cut in the darkness or of traps rifled in an afternoon of fog.

Sarah Holt could always set her straight again, redeem their hemmed-in world, rediscover its compensations.

"Don't expect so much of folks, Lucy, and especially of seafaring folks. The sea's a rough master. It brings out the best a man has, and yet it has a queer way of nourishing the worst. People talk about the great old days of sail, and they were great days, there's no mistake about that; but even then the sea could make rascals as well as heroes, and often a combination of both. I've spent my life on the sea or near it, and I'm still baffled by its ways. I've learned one thing, though, and that is that there's no terror like the terror the sea can strike into one."

"But not always," Lucy said. "Only sometimes, on strange days like this."

"Nothing's for always, thank God!" Sarah Holt said. "Let's make a cup of tea."

Lucy always made the tea, strong and black. They drank it from frail white cups with tiny green sprigs of leaves, which had come from the West Indies a century ago, brought home to Shag Island by Sarah's father.

"I used to see that terror in those years aboard ship, but I never grew quite used to it. When we'd have weeks of running before the Trades, with everything fair and not so much as a change in sail for days, no sailor could imagine a better life. But let us sit for weeks in the doldrums or work for more weeks against head winds and storms off Cape Horn, and this awful, sickening fear began to clutch everyone aboard. It wasn't the fear of never getting a wind again, or of being dashed to pieces on some reef, or of capsizing in mid-ocean. It wasn't heat or cold or drowning that men were scared of. It was just of being alone and lost in your mind in all that immensity of water over which you had no control. I used to watch it work. You got confused at first and then afraid, and after you'd been afraid for days, you got angry, and then you became dangerous to yourself and everybody else.

"I remember a sailor whom we picked up once after we'd made Mauritius on a long China voyage. He was a Scotch fellow from those islands off the west coast, just an ordinary seaman given to rolling from port to port as so many of them used to do. They say the Scottish people are stern and somber by nature, but this boy was the merriest boy you could ever imagine. He played jigs on an old, tuneless fiddle he had, stowed away in his gear, and got everyone dancing in their off-watches. My husband said he was the best sailor he remembered in a hundred crews. He would scramble up the rigging in the worst of weather and sing even while he was lying up there across the yards. There was nothing he couldn't do from patching canvas to giving the cook a hand in the galley. We were all plain grateful for him until we struck an awful calm in the Indian Ocean, just miles upon miles of glaring sea with the wicked sun beating down, and men stripped of most they had on and sleeping on the decks because the forecastle was a furnace.

"After about a week of it, when everybody was getting on edge, this boy just went clean out of his mind. You could see him getting more and more mixed up in his thoughts, and then surly and mean. One afternoon in the dog-watch, when everyone was tense and ill-tempered, and the sun and calm seemed to be mocking us all, he dashed below and came back with a knife and began to scare everybody out of their wits. He had slashed the first officer badly and a couple of seamen besides before they got him tied up. They bound him to a capstan, for he'd have died below decks in the awful heat. He just sprawled there, staring at the sea beyond the quarterdeck. I've never seen such fear in anyone's eyes; and I've seen plenty.

"I'd just been finishing a shirt for my husband, a nice one with fine tucking across the front, the kind shipmasters wore in those days when we sometimes went aboard other vessels for dinner in various ports of call; and because I couldn't bear to see him there with that awful look in his eyes, I took it across to him and said I'd made it for him and would keep it for him until we reached Hongkong. He didn't look at me or at the shirt at all. He just began to cry like a little child, sobbing for what seemed hours until he was fairly worn out. I'll never forget those sobs in that motionless air with the sun a great red ball on the horizon. The bos'n, a kind man from over on Shag Island, who sailed with us for years, stood by in that awful heat and kept wiping the tears and sweat from that boy's face. And after he was quiet and the bos'n had fed him his supper, he asked for his fiddle. I was proud of my husband when he decided to take a chance and let him have it."

"Did he play a jig on it?" Lucy asked, staring into the fog and slanting, wind-driven rain. She was for once unmindful of the store and of Joel's mistakes in adding. She knew she would never forget that boy sobbing above all those miles of glassy sea; and she thought she could not bear it if he had not been allowed to play a jig.

"He did, the merriest one he'd ever played, and we all danced to it. My husband and I led off, and everybody joined in. We forgot the heat. We danced and danced with the bos'n calling out the changes. Even the first officer danced with his bandaged hands. And when we were all fit to fall on the deck, the first stars came out, and we caught a whiff of wind off our port side and saw some ropes aloft begin to swing."

She paused, and Lucy folded up her work. The old clock from London struck four echoing tones. Fishermen ate their suppers at five.

"I'm not saying the fear's just the same in places like this," Sarah Holt said, "but the sea works in much the same way on people wherever they are if they have close dealings with it. It's out there waiting to make or to break you. And after it's been breaking you for weeks on end with fog mulls or contrary winds and giving you back little or nothing for all your work, you get lonesome and afraid and bitter. Men, even in outlying backwaters and on the islands, don't cut trap-lines or steal lobsters because they're mean by nature all the time, at least most men don't. They just get panicky, and then the worst in them beats the best."

6

The others of Sarah Holt's neighbors, except perhaps Sam Parker, never knew her as Lucy Norton did. Less thoughtful than Lucy and absorbed by the daily details of their way of life, handed down to them by tradition or bestowed upon them by necessity, they did not ponder upon their springs of action or even clearly recognize them. They would have been astonished and perplexed had they realized for a moment how well this old woman knew them and with what charity she understood them. But they were never unaware of her, sitting there in her house above the cove and even in her last days scanning the sea from her front porch. They would have pitied another who had had to cope with privations unknown in years gone by, with a son like Thaddeus and with all the trouble and shame which he had caused his mother; yet in a queer way, they realized, she did not welcome pity. Sometimes they found themselves thinking a bit uneasily about her, of what went on in her mind, and of just what made her different from anyone else they knew.

When she had died, they found themselves almost unable to imagine it.

Hannah and
Benjamin Stevens

❖ ─────────────────────────

◆ ─────────────────────────────

IN THE earliest years of the present century, before he embraced the fisherman's life, Benjamin Stevens had been a lighthouse keeper. This calling, always an honored one and often handed down from father to son before the Coast Guard took over the maintenance and service of light-stations, had been in the Stevens family for three generations. Ben was born and reared on less than an acre of rock, which, rising from the open sea seven miles from the mainland, marked one of the most dangerous of channels. His open-air playground as a boy had been a precarious one, and his paternal roof-tree a small gray house attached to the light-shaft which rose a hundred feet above it.

From his childhood he had climbed with his father the narrow, winding iron stairs which went up and up to the lantern-room and watched the trimming, cleaning, and filling of the great oil lamp, which was raised and lowered every four hours from twilight to clear dawn. Not an especially sensitive or imaginative child and knowing little or nothing of the activities of other children with which to compare his own, he took his singular background and experiences very much for granted. It was annoying, but not particularly strange to him to be securely tethered at high tide to an iron rung of the outside ladder which mounted the tower, until his mother decided that he was old enough to take his chances.

At low tide, when his playground, at least on fair days, extended some twenty feet on all sides, he found plenty to do. He could scare the great flocks of terns, cormorants, and gulls from their perches and enjoy their raucous, angry cries; watch the seals, which with slow, muscular writhings pulled their sleek bodies awkwardly from the sea to the crests of the ledges and boulders to lie there in the sun; search among the olive-green masses of kelp and seaweed at the water's edge for drift cast up by the tides, lobster traps and buoys, cork floats, clear glass toggles, lengths of tangled rope; and explore the deep fissures which the pounding, gigantic seas had cut among the high, jagged shores. At any time, except on the worst days, he could fish from the end of the slip leading from the boat-shed; and he always relished the excitement of the unannounced visits of the Government cutter, loaded with oil and other supplies and bent on inspection. During the summer, when the sea allowed a landing at the slip for the tenders of sloops and launches, visitors sometimes came to climb the many stairs to the top of the tower, see the workings of the great lantern, look out over the wide view of islands and of distant coastline, and exclaim over the desolation of such a life. As he grew older, he set a few traps off the sloping ledges on the west side of the station and caught some lobsters for the family table or even for his father to carry on his infrequent trips to the mainland after mail and more supplies. His only playmate was his sister, five years younger than he, whom his mother worried about and who was often an annoying hindrance to his plans and projects.

His schooling was scanty enough, confined to the little help his mother could offer with books lent by the nearest mainland school and to the fortnight once or twice a year when the Seacoast Mission boat brought the light-teacher, a dauntless young woman, who for eleven months out of the twelve was carried from lighthouse to lighthouse where there were children and taught whatever she could in that brief space of time. Her school-room with them was the circular entrance to the shaft, a tiny room furnished at these intervals with two small desks and chairs and a pine table for the teacher, whom his little sister always welcomed with far more eagerness than he did. When he was twelve, his father left him on the mainland with relatives there for three or four terms of school. He did not like books any better there and felt out-of-place in the unfamiliar surroundings.

He grew to be a big young man, tall and heavy with powerful muscles and great strength in his hands, though he was quick and even agile in all his movements. He was handsome, too, after a rough, masterful manner, with a kind of arrogance assumed to disguise his uncouthness. He took it for granted that he also would tend a light, probably the same one, when his father was through and had gone to the mainland to retire after his ceaseless struggle to keep the channel safe for passing ships of every sort, whether fishing-boats, lumber schooners, coastwise steamers, or pleasure craft.

2

As he grew older and began to feel uneasy stirrings within him and to think, slowly and practically, about other necessary details concerning his future, he now and again, when tides and winds permitted, crossed by himself to small mainland towns to see the sights, and to take in a dance, once he had summoned up the courage. Sometimes he met there the son of a neighboring light-keeper in an equally dangerous stretch of sea some ten miles away. The two of them not infrequently picked up an argument or even a fight with mainland boys, in which encounters they easily came off victorious and at the same time established a reputation for both brawn and prowess.

On one of these trips and at the close of such a combat he met a girl named Hannah Alley. Her father owned a clam-factory and was a man both of parts and of money in their coast village. Hannah was a slight, pretty girl, who danced well, was willing to tolerate his stumbling, uncertain steps and even to guide them, and who quite clearly admired him. She wore nice clothes which she had actually bought in Boston during her yearly journeys there with one of her friends. She was a good girl, too, everyone said, cheerfully worked for her father in his factory, and played the organ in the Baptist church. Her parents did not much favor her dancing; but, as young Benjamin early concluded, she held the upper hand over them both. He had never been admired by a girl before; and since things were

as they were and his certain future required stability and
assistance, he rather roughly one evening asked Hannah to
marry him. To his immense relief and satisfaction, she said
that she would. He was twenty-one at the time, and Hannah
was going on twenty-four.

Things worked out admirably for them both, he often
thought rather ruthlessly, though that manner of his thinking never entered his mind. His father at fifty was stiff
from rheumatism and found the inescapable stairs to the
lantern-room more and more difficult; and his mother,
deprived from her girlhood of mainland ways, longed to
recapture them and live for the rest of her life away from
impenetrable fogs, tumultuous seas, constant anxiety, and
the lonely screaming of gulls. His sister, far more advanced
than he in their uncertain, slender opportunities for study,
could now go to high school as she had long dreamed. What
could be more fortunate for all than that Ben should take
over the rocky reef? He would be the third of his stock to
do so. Government inspectors and officials had long since
learned that family tradition and upbringing were safe
qualities to depend upon for the guarding and guiding of
vessels.

After a year on the light-station, when the admiration
of her friends had worn as threadbare as her own earlier
sense of adventure and daring, even of heroism, Hannah
became unhappy and homesick. She was often petulant and
demanding, wanting more trips to the mainland than Benjamin, knowing the unpredictable ways of the Government
cutter and honestly feeling his new responsibility, felt safe
or fair to make. She was afraid in storms, and superstitious
over sudden ominous happenings, Northern Lights, shoot-

ing-stars, sun-dogs, strange rings around the moon. When her baby was coming and she felt lumpish and unwieldy in her small kitchen, which with the rest of the house must be kept immaculate, every dish shining and in place, every brass doorknob polished, no matter how she might feel, she insisted upon being taken well ahead of time to her father's home. She would not depend, she said, on any doctor reaching the station in time, even if weather conditions were possible to fetch him as they often were not. If the wives of other light-keepers chose to wait out their time in such frightening solitude, they were quite welcome to do so.

Their sojourn at the light-station turned out to be a brief one. Hannah found it no place for a baby and a perilous one for an irresponsible, restive child. She hated tying her little girl out of reach of danger and was forever expostulating against it to her husband, who grew steadily more impatient of her complaining discontent. One spring day, while he was busy about the shore and boat-shed and she was setting things to rights within the house, the child, then three years old, fell into a deep fissure in the rocks. They did not find her until an hour later, after the tide had turned. Whether she had been killed by the fall or drowned in the rising waters, they never knew.

Just as soon as another keeper could be found, Benjamin Stevens joined Hannah on the mainland. Not many months afterward he became a lobster fisherman in the settlement at the cove. For several years they were haunted by the memory of the child, although they rarely mentioned it to each other and never to anyone else. Hannah felt precisely as though a jagged rock lay somewhere deep within her, turning restlessly from time to time as rocks are turned

(95)

about by the incoming tide. When a son was born to her, she felt better able to forget the restless jagged rock which had for so long torn at her and weighted her down.

But it had never forgotten her.

3

From the beginning of their life in the fishing-village they maintained more interests outside it than did their neighbors. Ben, Hannah said, welcomed a change once in a while and surely warranted it with all his hard work. He was a steady, careful fisherman, although never an adventuresome one. He did not invest his hard-earned money in a herring weir to supplement his lobster traps, partly because he did not like working with anyone in the association which a weir almost demanded, largely because he was wary of any investment which in the beginning was likely to require capital from one or another of the large fish-packing concerns eastward. He preferred to stick to lobsters and even to sink his traps in relatively quiet waters. But with the death of Hannah's father and her considerable share in the profits which he left behind him, they found themselves, some ten years after they had come to the cove, in comfortable circumstances. Long before Joel Norton had saved enough money to buy his truck, Benjamin and Hannah were amply able to afford a car. They used this, especially on Sundays, to escape their humdrum existence and to see something of the world outside it.

To them this world outside meant primarily a small church some twenty miles inland at the head of the Tidal River. It belonged to none of those sects common to New England coast regions, but was instead a somewhat lonely outpost of extreme evangelism. Its rather diminutive congregation consisted of ardent, anxious people who adhered unquestioningly to the inspired Word of God as revealed to His prophets and to His apostles. In their minds every sentence of this Word to its utmost syllable had been dictated by God Himself and proclaimed once and for all time by His Son. Their theology, if their simple tenets could be dignified by such a term, was extremely tangible. It demanded an entire recognition of sin, past and present; humble repentance and public acknowledgment; and total immersion in the cold waters of some bay or estuary in the sight of brethren and sisters, after which salvation was unconditionally guaranteed. In comparison with its unrelenting demands, even the rigid doctrines and practices of Hannah's earlier Baptist faith were expansive. Indeed, upon her entrance into this newer communion of saints, she had been made so conscious of guilt and remorse that she had submitted not only willingly, but eagerly to a second baptism.

Benjamin's conversion came some twenty years after he had settled down as a fisherman. At first he had taken Hannah to church on Sundays, and occasionally to the frequent evening prayer-circles because things went decidedly better at home if he allowed her these satisfactions. But gradually the sense of belonging somewhere, of having something to hang on to, began to stir some unsatisfied need within him also. Perhaps the sentiment often uttered by minister and congregation, that, just as the Lord had chosen fishermen in the beginning to carry on His work, so now He needed them above all others to conclude it at a time which all signs proved to be close at hand, had its part in his slow awakening. He had a good and true bass voice and, after some time of uneasy embarrassment, began to sing on Sunday mornings with the others, while Hannah, pumping the small organ with her feet and pressing the often recalcitrant keys, sent him glances of encouragement from the platform to the right of the pulpit. The grateful, admiring comments of his new neighbors at the close of the morning service began to sound sweet in his ears. All these influences were, in fact, far more powerful than Hannah's anxious warnings about the Ultimate Future when families would be either ruthlessly and justly separated for eternity or reunited to live in bliss, all earthly imperfections blotted out, all sins forgotten. When at last during one of the frequent revival seasons he summoned up the courage to walk forward to the mercy-seat, his great head and shoulders towering above the few weeping, shuffling figures of other penitents, he felt relieved and settled in his mind, even without that strange inner peace which the preacher had so eloquently promised him.

He was never a power in the church like Hannah, who, from her entrance into it, had begun to mould its destinies. Her position as organist and as superintendent of the tiny Sunday-school in themselves gave her prominence; but these were but outward and visible signs of her strength and influence. When Ben had left at dawn to haul his traps, she always sat down at her kitchen table for another cup of coffee over her daily Bible reading. It was then that she pondered over less obvious church problems, stealing into their intricacies like some black otter making its way into hidden crevices among the rocks and ledges of the shore. Secure in the Scriptural decrees that each is responsible for the sins and shortcomings of his brethren and that unworthy branches of the tree must be hacked down, cast into the fire, and burned, she carefully considered those human beings who together made up the congregation, weighed each in her self-constructed balance, and frequently found several wanting. Then, later on in the day, as she wove the tight meshes of her husband's bait-bags and trap-heads, her mind set to work to see what best could, and, indeed, must be done for the good of all.

When she was not employed in her religious employments and maneuverings, Hannah was a kind neighbor. She was a natural and excellent nurse and would willingly spend hours in caring for anyone who was ill or had been hurt, often driving miles to do so. A superlative housekeeper and cook, she loved bestowing products of her culinary art upon all her fellow housewives, not alone on those less gifted than she. She rarely imposed her religious convictions upon others in her immediate neighborhood, largely because her easily embarrassed husband had forbidden her to do so.

"Keep your mouth shut!" he said. "Let them all fry in Hell if that's what they want to do."

Although such a summary injunction seemed on the most cursory examination directly opposed both in word and in spirit to that missionary zeal encouraged every Sunday by the shepherd of their flock, Hannah wisely concluded to follow it.

4

On the day of old Mrs. Holt's funeral, when she went to keep the store during Lucy Norton's morning absence from it, she took with her in the bottom of her work-basket some sheets of paper upon which to write a letter. The paper was placed within the pages of her Bible which she also carried, both because she felt more secure with it close at hand and also because she might well wish to quote from it. The letter must be written with the utmost circumspection and care; and although for many days she had been composing it in her mind, she had not yet arrived at the wisest and most effective manner of its expression. She had decided upon the store as the best possible place for its composition since there could be no trade to speak of with everybody hauling traps and even the children safely, or unsafely, out of the way, and since for some completely inexplicable and vaguely troublesome reason, she could not seem to get it down on paper among the familiar sights of home.

For all her correspondence Hannah used stationery printed at the top with Scriptural messages and warnings. These sheets were purchased at the church, the profits from their sale going to increase the slender funds for missionary work. *For God so loved the world that He gave His Only Begotten Son,* one said. *Come unto me all ye that labor and are heavy laden,* pleaded another. And a third, *Though your sins be as scarlet, they shall be as white as snow; though they be red like crimson, they shall be as wool.*

This letter, which had to do with the present minister of her church, was addressed to the head of some Home Missionary Board in Boston, a board which presided over the placing of those pastors dedicated to the obscure and zealous sect, here represented by the Tidal River congregation. Hannah had already spent several weeks in earnest

thought over the minister's works and ways and a lesser
time in what, she had convinced herself, was equally earnest
prayer for guidance. It was difficult to write the letter, for
only a few months previously she had reported, as church
clerk, to the same Board that all were pleased with young
Mr. Simpson, his wife and three small children, that souls
were being miraculously saved under his inspired preaching,
and that the parish was on the point of raising his meager
salary after his four years of devoted service among them.
Now that she had a far different report to make, she
must seek carefully and prayerfully both for right words
and for valid reasons which would bear weight with the
Board in Boston.

Mr. Simpson was an earnest, fervent, ungrammatical
young man of twenty-eight, who had early been converted
in a Gospel Tent, raised along some remote Damascus
Road among the mountains of Kentucky. At about the same
time and under the same circumstances he had found his
wife, a shy, frightened girl, who looked upon him as a
veritable major prophet and pathetically wanted only to
contribute her all, which was little enough, to his advance-
ment. He honestly and quite humbly believed that he had
been specifically chosen to rescue his fellow men from the
perils of sin; and once he had discovered a struggling
training-school which cared less for learning than for zeal
and was in sad need of recruits, he had spent two years in
the study of the Bible and in the revivalistic methods of its
interpretation. The congregation at the head of the Tidal
River was his first parish; and he and his wife had transported their few shabby possessions with a pride and
pleasure which, he sometimes feared, competed dangerously,

perhaps even wickedly, with the purposes of God for him.

From the beginning of his ministry and even, so far as he was aware, until now, Hannah Stevens had been his bulwark, the strong, defended wall and tower of the ancient prophet. She had manned his Sunday-school, played his organ, suggested subjects for his sermons, over which he laboriously struggled, and ferreted out needy souls in a dozen isolated hamlets. She had encouraged his wife with confidence and new recipes, replenished their depleted larder, and clothed their children. He had, it is true, felt somewhat diffident with her husband, whose very size and strength were overpowering. Still, Benjamin Stevens was conspicuously generous with his Sunday morning offerings; and once, on a Saturday afternoon, he had taken the whole Simpson family on a sail in his fishing-boat, an experience which so excited them all that they had found it difficult on their return home to compose themselves for family prayers. The minister thanked God daily for Hannah, who was in truth his Dorcas, Priscilla, Lois, and Eunice combined in one woman; and he prayed almost desperately that he might be allowed to remain permanently in this particular vineyard where the compensations were so many and the need, likewise, so great.

Hannah, in her turn, as she strove now to write her letter, recalled uncomfortably these more pleasant aspects of Mr. Simpson's stay among them. She strove, too, to rid her mind of other tenacious memories: of the reckless ruin of her petunia bed by the oldest of his children; of the able substitution of little Mrs. Simpson at the organ when snow had made the long road impassable for Benjamin and her. Yet these, she told herself with ever-growing conviction, had nothing to do with her real and honest conclusion that the congregation needed a change in pastors. It was, of course, undeniably true that a minister's children should be an example rather than a warning and that a minister's wife should not be apparently eager to take over duties so well performed by others. Yet these were only straws in the wind, and she would not for a moment allow them to influence those indisputable facts supporting Mr. Simpson's incompetency, facts which she, sitting behind the counter in the store, was striving to assemble in her mind and to write down on her paper.

She decided finally that she might well begin with his lack of courage in dealing with intemperance, alas! so common an evil in homes quite within his reach, now that the church had provided him with a second-hand car. She remembered that there were fortunately among her sheets of note-paper some appropriately headed: *Look not thou upon the wine when it is red. At the last it biteth like a serpent, and stingeth like an adder.* Just then, when she had barely determined upon her opening words, the Randall child came in for ten cents' worth of licorice, the other children standing in a silent group outside. Once she had put the ten sticks in a paper bag before the child's frightened stare and called from the porch a warning to her grandchildren not to forget for one moment what she had told them at breakfast, she returned to her letter.

"Rev. Sir," she wrote on a piece of tablet paper with a pencil, for the letter must be composed to her satisfaction before it was transferred in ink upon the proper sheets.

"Rev. Sir:

As clerk of our church, and organist for ten years,

and Supt. of our Sunday School for many more years, I feel it is my painful duty as a Christian to set before you . . ."

5

Even while she was setting down the first words, she saw the children trailing up the hill in search of flowers for old Mrs. Holt. Why the sight of them setting forth should so interpose itself between her and her task, she could not see, and she felt suddenly flushed and irritated. Then, when she had banished the children from her mind, she recalled the disturbing fact that she had not been asked to help in performing the last duties for Sarah Holt as she had been asked to do in numberless similar situations and in far more important communities than this one. She thought of Ben out there hauling his traps and of what he might say unless she could keep her letter concealed from his knowledge. She even recalled, in spite of herself, the surprised and delighted exclamations of the minister's wife over cakes and pies she had made for her in the recent past, one cake identical with that which she had baked this morning in an effort to quiet the resentment and suspicion which nagged at her from more than one quarter. And behind and through all these unwelcome images and recollections which seemed maliciously bent on tormenting her, lay the face of the dead woman, which she had purposely not gone over to look upon.

Lucy Norton's clock on the shelf above the stove ticked away the two hours upon which she had counted for her letter. There was still no question in her mind that it must and would be written; but today was clearly not the day for it. When she saw Lucy coming down the road, she put away the sheets of paper in her Bible and concealed both in the bottom of her workbasket before she went out on the porch.

Nora and Seth Blodgett

◆ ───────────────────────────────

S ETH BLODGETT had been among the
fishermen waiting at the store for the news
which Lucy Norton brought when she
returned at midnight from the Holt house. He was not
young any longer and rarely kept such late hours; but he
had felt that respect for old Mrs. Holt demanded his waiting
up. He was a tall, stooped man with thick gray hair and
large brown eyes, now vacant of expression for he was fast
becoming blind. Lucy often thought how handsome he
must have been when he was a boy.

When he had made his way to the door with the others,
Sam Parker guided him down the steps in the rain and wind
and walked with him to his house which stood up the road
a piece next to that of Benjamin Stevens. Ben had gone on
ahead since Hannah would be getting fidgety over his late-
ness. They could hear the sound of his heavy boots like dull
thumps beneath the fog whenever the pounding of the
waves of the beach allowed any other sound.

2

Nora Blodgett was fidgety, too, though in a far different way from Hannah Stevens. She lay in bed, well over on her side of it, and wondered what Seth would say, if he said anything at all, when he came in, and, if he did say anything, what she might just possibly say in reply. For a steadily increasing number of years now she had made up dialogues between them, partly from fading memories, largely from hope. The one tonight should take place after this manner:

"Has Sarah Holt gone, Seth?"

"Yes, Nora. She died about an hour ago."

"Did she die easy, without any pain?"

"I s'pose so. Lucy didn't say anything to the contrary."

"She was a wonderful old woman."

"Yes, she was all of that."

"We're going to miss her terribly around here."

"We surely are. There's no mistake about that."

"We shan't see her like again, in places like this."

"No, we shan't, more's the pity."

"Would you like a cup of tea, Seth? You must be tired out."

"Well, I must say, one would taste first-rate if you're not dead beat yourself."

Even as she framed the words, saying them over to herself in the shadow of the kerosene lamp on the table, she knew they would never be spoken. Not only the most usual occurrences, but the most extraordinary of happenings, a major run of fish, Carlton's new boat, the coming of the Randalls, had failed to make her improvised dialogues take audible form so that they might break down this impenetrable, lonely wall of silence between her and Seth. Whatever had she, then, to hope for from Sarah Holt's death?

Nor would words alone suffice, she realized, even if they could be uttered. Conversation meant more than words. It meant sympathy and concern evident in the tone of voice, in the inflection of speech. It had been years now since she had been able to keep petulance and vexation out of her voice when she spoke the most necessary and inescapable words to Seth. She felt neither of these emotions, she told herself, wiping her eyes with the hem of the sheet. Their semblances were masks for her real feelings, and just where they had come from she did not know. She hated them and despised herself because of them, strove to conquer them, but they defeated her with their whining accents and irritable undertones. She was, instead, filled with pity, with desperation for the present, with terrors for the future. Yet pity and terror had not availed to keep the wall between them from rising higher, built of the hard gray stones of mutual embarrassment, of the loss of youth and hope, of the exhausting rigor of toil. It had grown imperceptibly at first, yet relentlessly, until now either to shatter or to scale it was inconceivable and visualized only in ironic moments of fancy.

She lay now in the small, shadowy room and listened to the rollers crashing upon the high beach from the full, wind-driven tide. With every roar the lamp shook and flickered. It was well past midnight, and after but a few hours of sleep another gray, wet dawn would break over the dank, empty flats. They would wait in silence for the sea to return so that they could get underway.

She heard the door open and Seth's uncertain steps in the kitchen, his hands fumbling at the light there, the sound of his breath as he blew it out. She gathered her body together like a diver before his plunge into deep, icy waters. She would try yet again, control her voice, begin, at least, to give birth to her imaginary dialogue.

Seth came into the room in his old shirt and corduroys, unhitching his galluses as he came, making ready to step from his trousers. She sat up in bed, trying to meet his eyes and then suddenly realizing that he could not return her gaze. This knowledge, bitter and cruel though it was, lent strength to her resolve.

"Is she gone?" she asked him.

"Yes," he said.

He got into bed. She blew out the light and lay down again, resisting an impulse to move even farther away. Once more she summoned back all her waning faith.

"You wouldn't want some tea, I s'pose," she said.

He was stunned at her words. For the fraction of a minute he was like a child who sees a soap-bubble and longs to grasp it before it bursts into nothingness before his eyes. Then he turned his back upon her and burrowed his head into his pillow.

"You gone crazy?" he said angrily. "It's nigh on one o'clock. I want to sleep."

3

Before he came to the cove village as a lobster and herring fisherman, Seth Blodgett had sailed to Georges Bank and the Bay of Fundy for cod or halibut or haddock, and at times, when unknown impulses sent them there, for countless thousands of blue and silver mackerel. It was in this grueling labor that his eyes began to trouble him. The bitter cold of night watches when his eyeballs seemed to congeal into staring circles of ice; the glaring sun which sometimes turned the limitless surface of the sea into liquid fire; and the haze of mingled fog and sun which blurred one's vision after hours of peering into it—these after twenty years took their toll. When he was nearing forty, he sold the schooner, of which he had been finally owner as well as skipper, and began to seek along eastern points and bays for an easier manner of life which might mitigate and postpone, if it could not entirely dispel, the fear which at intervals gripped and shook him.

Free now to give time and consideration to a home ashore which should be his and not his father's house, he felt himself fortunate beyond all expectation when he met and fell quite thoroughly in love with an upcountry girl named Nora Bartlett, who was working in a summer hotel just across the border in a New Brunswick coast town. She came from farming stock and had never known the sea. Could she have then foreseen how intimately she must come to know it, she might have hesitated even in the face of the charm, strength, and untutored persuasion of Seth Blodgett. She was ten years younger than he and frankly eager to marry and have a family.

The size of their family disappointed them both. Three years after their arrival at the cove their one child, a daughter, was born. They had proudly given her what the coast knew as "chances," four years at the nearest high school and two more in training to be a teacher. Neither had been displeased, however, (and her father secretly relieved) when she suddenly decided to marry a boy named Carlton Sawyer. He belonged to the crew of a seiner, and she had met him only a few times, whenever his compact, snub-nosed boat with its trailing dories made the cove harbor for some layovers at night.

The first ten years of married life left little to be desired for either Nora or Seth. His eyes improved or at least grew no worse, which was lucky, he liked to say, since he needed them to watch his wife. Let any craft slip inside, whether a sardine-boat, a Coast Guard cutter, or one of these new cabin cruisers, and every man who trudged to the store for supplies looked twice at Nora, given the chance. She was tall and straight then, undeniably pretty with shining blue eyes and easy, cordial ways. Her husband was inordinately proud of her.

4

In those earlier years Seth fished far out, off the ledges near the great light. He had saved money from his generous shares at Bank fishing and could afford a long string of traps. He also went into partnership with Daniel Thurston in his weir off the inside shores of Herring Head. Two or three good seasons in a row swelled his account in the Machias Savings Bank and added to the dirty, redolent rolls of bills which, like most fishermen, he kept stowed away in a canvas wallet, concealed now in one place, now in another, about the house. Once, in a hurry to start hauling, he forgot the new hiding-place and carried his fat wallet to sea with him, deep in the inside pocket of his leathern jacket. Nora sometimes remembered now how in the tossing swells and swirling fog she had cried to him

from the stern thwart while he tinkered with his stubborn
engine:

"Suppose she fails you, and you get stove up and
drowned off those ledges with all that money on you?"

And how he had laughed above the rising wind and tide
and called back:

"You're along, ain't you? We'll spend every damn cent
wherever we go!"

She had hauled with him from the start except for the
years when Mary was too little. She had loved their getting
up before dawn, eating doughnuts washed down with cups
of strong, heavily-sweetened coffee, pulling in the dory or
wading out to it over the watery flats, boarding the fishing-
boat, getting underway in the early light, skirting the
moorings and the rocking bell-buoy, seeing on clear mornings in the distance, beyond the headland on one side and
Shag Island on the other, the open, rolling seas through
which they would cut their sure, V-shaped way. Although
it was hard even to imagine it now, Seth often, just as the
sun burst above the horizon, slowed down the engine or
stopped it altogether while they tossed and rolled back
and forth along the swells of trough.

"I always like to sit quiet for just a minute while the sun
comes up," he had said. "I used to do it in the dories off the
Banks. It cost me some fish in those days and a heap of
twitting besides."

Her neighbors were generous in their comments about a
woman's hauling traps. She would pay for it later in a
dozen painful ways, not to mention the mistake of coddling
her husband..But she was a woman who always had the

future in her mind as well as the present, although her darkest visions of that future had been bright in the blackness of its reality. She learned every rudiment of lobster fishing and grew expert at each. She even kept her eyes on the engine though that was always Seth's province. These elements were simple and demanded little practice. The things which did demand practice and to which she never grew inured were those far less tangible elements within which they moved and worked: the recurring shock of the icy water in which the buoys floated and tossed; the fog closing over them, its density and danger prophesied by the moaning of the horn from the invisible light-station; the sudden squalls of rain which pierced her thick clothing to her skin; the lightning and thunder which, bursting upon them, mysteriously emphasized their defencelessness, their want of refuge and shelter; the sickening fear which smote her on days of high winds and heavy tides when they drew so close to the red ledges that they were struck by the backward lash of spray.

These apprehensions, at first only misgivings and forgotten when they had made the harbor, increased to terrors when she saw that Seth was beginning to peer at his engine, bent almost double over it, and noticed his hands moving uncertainly upon the traps as he emptied and rebaited them. And since in the face of her own silent dread she could not bring herself to free her anxieties to him and to plan with him for a different future, she found herself resorting to querulous expressions of irritation and anger. One day, when both wind and tide were against them and she was handling the gaff in the rocking boat, she lost it overboard in a sudden pitch. She was shocked when she heard herself replying to Seth's ill-concealed contempt and disgust over her ineptitude:

"Just remember, will you, that I don't have to be out here helping you!"

She knew then with a far more sickening feeling than ever a submerged, wave-swept ledge had given her that her impulsive, angry words would be repeated indefinitely and that they marked the beginning of a tragic end.

5

They set their traps now up the Tidal River; and since their fishing-grounds were nearer than those of their neighbors, they did not get underway so early in the morning. Nora had time to put the house to rights before she went down to the beach. Seth always made his way there by himself a good half-hour before she had things ready to leave.

The path beyond the road was a familiar one, past his fish-house, down through the nettles and the asters, on to the shale and the beach stones. Knowing every hollow and curve, he could traverse it easily, even with the awkward weight of the oars which he carried. Nora, glancing now and then from the front window at sunrise on the day of Sarah Holt's funeral, watched him doing it and understood his sullen pride. Then she found herself seized by exasperation and angry impatience when, stumbling suddenly in the heavy gravel, he lost his hold upon his oars and had to fumble about among the stones on his hands and knees before he got them together again.

As she pulled on her rubber overshoes and got into her

old sweaters, for even on this unaccountably fair morning it would be cold on the water, she wondered as she had done for fully a thousand times just how this daily nightmare would come to a close. She now managed everything about the boat except for the engine. She had grown quick and skillful at the tiller, sidling up to the green buoys, seizing their ropes, placing these in Seth's hands so he could haul in the heavy traps. He would never yield the engine to her. He managed somehow by touch rather than by sight to work its familiar controls.

Perhaps it would end, as actual nightmares often did, with a sudden start. Only instead of being shaken awake and finding oneself safe in bed, they would find themselves with a dead engine and drifting hopelessly away from the inner reaches of the Tidal River toward the turbulent seas which broke against the rocks and shoals of Shag Island. Their particular outpost of coastline, as she well knew, was generous in its records of such disasters.

What it could not so tangibly record were those ancient, more tragic disasters of estrangement and isolation, in which human beings, torn with panic and pity, turned desperately, even in their reluctance and their shame, to cruel recriminations and anger, or to more cruel silence.

6

She could not recall so beautiful a day in all her years at the cove, she thought, as, standing at the tiller, she cautiously watched their passage through the wide mouth of the Tidal River against the swiftly ebbing tide. The rush of the sea at the northern end of Shag Island was barely perceptible against the cliffs and ledges. Even the small islands,

Hardtack, Pumpkin, the Castle, Eagle Rock, lying eastward beyond the protective neck of land which, with their own more precipitous point, lent designation to the Tidal River, were scarcely ringed with foam. She watched Seth bent over his engine. She wished with all the longing she possessed that she dared to venture an easy, pleasant comment upon this incredible change in weather, this sudden lifting of fog and stilling of wind.

Just as they were passing the Holt house in its high field on their left, she saw Thaddeus Holt walk from the back door to the barn, which served him as his fish-house, and begin to shift some buoys stacked against it. The heavy, blunt sounds as he repiled them echoed through the motionless air. She wondered whether or not she should wave to him on the day of his mother's funeral. He answered her question by waving his arm in greeting, and she returned his gesture with an odd, pleasing sense of companionship and gratitude.

She had hardly done so when she was astounded by a question from Seth.

"What's all the racket ashore?" he asked with no sign of irritation in his voice.

She thought for an instant that she could not answer because of the quick tightening of her throat.

"It's Thaddeus," she managed to say. Her voice was quiet and even. "He's shifting his buoys."

A heron flew above them, the sun revealing the blue of its wide wings. She thought, looking at it through her tears, that she had never seen anything more lovely. She owed it something. She said:

"He waved to us." And then, "He's watching us now."

She saw Seth get up from his engine-box and face the shore which he could not see. She watched him raise his arm toward Thaddeus. I can't cry, she thought. He'll hear me. He hears everything. Again she controlled her voice.

"He waved again," she said. "To you."

He sat down, bending over the engine, fussing at the controls. He cleared his throat and turned away his head. Then he fumbled in his pocket for his pipe. He did not light it, only clenched his teeth about the stem. After a few minutes, as they moved onward toward the first of their buoys, she knew that he was about to speak again. She could almost hear the confused murmur of the words struggling for utterance.

"He's a good man, Thaddeus is," he said, spacing the words slowly, striving to sound each without embarrassment.

She could not expect any more, she thought, but more came.

"With all his carrying on, he's a better man than most of us."

She wiped her eyes with her sleeve. She could hardly see the green buoys ahead floating on the calm surface of the water. She became angry, but only with herself. She was out here to haul traps with her husband, not to cry, like some silly child.

"He is a good man," she said, her voice breaking a little in spite of her efforts to master it. "It'll be lonesome for him now."

"Most likely," he said.

A flight of ducks rose from some hidden cover on the shore, making seaward with a whir of wings.

"Ducks," he said quietly, biting on his pipe. "This was always a prime place for ducks. Don't you remember how we used to come up here after them, years ago?"

Mary and Carlton Sawyer

✧ ─────────────────────────

E VERYONE who had known Mary Blodgett at the Normal School said that she
must be quite out of her mind to marry
anyone at all, if it meant settling down in that lonesome
cove on that God-forsaken point of land with all *her* gifts
and possibilities. Since she had escaped from its privations
and exigencies, whyever should she want to return to them?
Perhaps they would not have been so vehement in their
condemnations had they known young Carlton Sawyer.

Carlton Sawyer came from the region of Bath where
members of his family had been known for nearly two
centuries as the builders and owners of ships and as the
masters of them, too, in the foreign trade. Had he lived
a century earlier, he would have without doubt commanded
them also, approaching China by way of Good Hope and
the East Indies and bringing home pepper and tea. Or he
might have beaten around Cape Horn on the westward
passage, bartering on his way with a dozen Pacific islands
from the Fijis to Hawaii before he reached Canton. One

could easily imagine Carlton as especially created for just such a life. The trouble was that the life had gone, past all recall.

Yet the sea was what it had always been, and he was born and bred to it, both by inheritance and by nature. Unlike his brothers he could not conceive of substituting a course in marine engineering, or navigation, or ship design for an intimate knowledge of its ways and dangers. Nor could he see himself, whatever his propitious chances, rising to a position of trust and security in the still active ship-yards at the mouth of the Kennebec. Once he had finished high school and refused college in the face of all inducements, he began at once to acquire that intimate knowledge, with the sea itself and the various occupations which it fostered and threatened as his only teachers. His family thought him an impossible rolling-stone, doomed to insignificance, although even they could not visualize him as a failure.

It was, perhaps, his attitude of gay irresponsibility, of taking chances, of giving all manner of watery pursuits a try, of laughing at Fate and Fortune, which appealed most strongly to Mary Blodgett during her two summers at home from the Normal School. As she saw her father and mother growing more withdrawn and bitter and felt herself being irresistibly caught in the intricate net of their hopelessness, she welcomed his infrequent appearances with an almost intoxicating relief. He was seining at the time, following a year of fishing with a Gloucester fleet and a nearer one of carrying lumber from the Maritimes southward. It was all in the day's work, he said. He was learning what he wanted to know. The seiner was a filthy tub enough, and he got a bit sick of sliding in his dory into bays and backwaters at night looking for herring with a torch, or of spreading nets at dawn along some low-lying shore only to take nothing or to break his back at ladling thousands of slippery fish into a stinking hold. When he knew enough about seining, he would learn the ways of weir-fishing and lobstering. Then perhaps he would master drag-nets. There was plenty of time, and everything yielded its excitements for him.

Everyone looked with admiration at his tall, spare figure, his shock of fair hair, and his steady gray eyes; and everyone liked him. Even the weir fishermen, the bitter rivals of all seiners, succumbed to his charm. When his boat made the cove for a night, and he had had a swim and then got cleaned up in one of the small shacks on the hill, he would spend an hour in the store, meeting the silent scowls of the weir-workers with great good nature, cursing with them the wretched ways of herring, assuring them that luck was seemingly against the men who dragged the shore-nets just as it was against them who dropped them from weir-poles. He could break down the reserve and suspicion of anyone, even of Seth Blodgett, with whom he fell into the habit of walking homeward to swap tales of catches off the Banks. Like his ancestors on their clippers and down-easters, he found eager girls in every port, although the ports themselves in these present days hardly savored of old romance. Once he had met Mary Blodgett, however, he more than half believed that there might be something in this notion of the one and only girl in the world.

This illusion had never once failed him during the ten

years of their marriage. Instead, its wavering lines had sharpened and deepened into those of firm reality. He had spent more years at lobster fishing than he had once planned, to be sure; but now that he had a new incentive, his future had taken on a different pattern. He decided that he liked a secure anchorage, at least for some time longer; and now that they had managed the new boat, he would not have to spend four or five months of every year puttering over traps in his fish-house or going into the woods for some lumbering to fill up time and eke out money. The *Mary Blodgett* was framed for winter fishing. There was nothing he might not do in her, no winds or seas she could not take.

Next to Mary, this new boat was his passion, sometimes competing, she claimed, with his two children for his care and devotion. When, safely out of sight of the other, and lesser, cove fishing craft, he really let her out and saw how she cut the water and how casually she rode the slopes of the swells after a storm, he wanted for nothing. The thought that she was paid for, too, glowed in his mind like one of her own riding-lights when he came in late on some autumn night. In his single days, when he had been wedded only to the sea, it had been easy-come, easy-go so far as money was concerned. That slapdash existence was over and done with, his wife said. She had quite enough to trouble her without the harrying worry of unpaid bills.

2

After she had made this remark lightly and yet often enough so that Carlton said it was beginning to sound too much like some old proverb in his ears, she found herself mulling over in her mind how true it was. Even into her happiness there crept shadows, dim shapes which at first she could neither clearly discern nor drive away. When they persisted in spite of her contentment, she began to try to define them. Their essence, she concluded, was of dread, not of fear for their present, but of foreboding for their future.

She did not fear for Carlton's safety, that common fear of most women whose husbands follow the sea. His overwhelming confidence had built up her assurance that he was proof against any perils from winds or tides or fog. Instead, she dreaded the days lurking in the coming years when even perils should have lost their insolence and the stoutest hearts no longer take triumphant pleasure in defying them. She dreaded what often seemed to be the inevitable results of the ceaseless routine of toil: the stealthy inroads of habit which had paid its toll to time and had long since lost its zest; the passing of youth and the sense of adventure and daring.

Whenever she saw her mother join her father on the beach for their endless daily round of labor, she saw herself twenty years older, defeated and silent. She saw her children wondering and troubled as she had been, storing up fears and uncertainties for their own futures. Impossible as these things seemed now, she knew that they waited, dark and threatening, in a hundred far-flung communities like their own. Whenever she and Carlton had taken a rare trip to town to shop and to see a movie, she always feared the turn from the highway onto the long, desolate road; for with the very turn a curtain dropped between light and music, plenty and security on the one side, and the unfulfilled longing for them, on the other. The few dark, old houses which they passed on the way home meant dull sleep after a day of struggle and labor; or the dim light in some kitchen window told of the uncompleted tasks of some woman, who herself had once been young and, perhaps, in love with the man she now took for granted and bore with as best she could.

Such creeping shadows, she felt sure, did not steal into Carlton's mind to darken so much as an hour of his days. Even in the long winter, which autumn by autumn she awaited with misgiving and dismay, he was buoyant, his occasional restlessness over forced inactivity not the restlessness of fear. Weather provoked, but did not dishearten him. He cursed the fog, but it was merely fog to him, not after many days an ominous prophecy or a symbol of all that was baffling, obscure and terrifying. Carlton, happily for her, did not deal in symbols or so much as recognize them.

And happily, too, there were symbols of quite a different nature to which she clung when the winter isolation crept relentlessly over the cove. In the early spring when the smelts rushed up the freed brooks in thousands on the high tides, they would get up in the dark and tramp for miles with their lanterns and baskets to some small tidal stream where they could gather a watery harvest in their bare hands. There would be summer afternoons, after hauling the traps, when they bundled the children on board the boat and swung across the channel eastward to some island where they would cook their supper, watch the moon rise, and sail homeward through its silvery path. The children were still too young to trouble her with the problems of their education and training, although Carlton saw his son as a Naval Commander, perhaps winning glory and medals in yet another war which gloomy people were forever prophesying. She planned no future for her daughter except secretly to determine that she should never become a fisherman's wife.

3

On the night before old Mrs. Holt's funeral she moved about the house, waiting for her husband to return from Shag Island with Sam Parker and Joel Norton. She had long since put the children to bed. She tried to read, but found it impossible, and then to tackle her heaped mending-basket, which likewise proved impossible.

Most of the trouble, she told herself, ashamed of her restlessness, lay in the fog. Even with the doors and the windows closed, it slipped through the jambs and under the sills. The lamp on the kitchen table was dimmed by it, the flame glowing uncertainly through a visible mist. When her head began to ache in the stuffy house and she opened the door for fresh air, sea-blown clouds of it swept past her to steal through the small rooms and settle in tiny beads of moisture on chairs and tables. Her hair was wet with them, and her sweater was clothed in a white film. The fog had so wrapped itself around the spruces behind the house and wreathed itself among the bushes before it that they gave forth a drizzling sound like a heavy, subdued hum. When she went into the bedroom to peer from the window at her father's house, she could discern no glow of light there or make out even the outlines of the house itself. Everything had disappeared into this enveloping white shroud, and, except for the dripping undertone from the trees, there was no sound. The tide was far out and silent. She would have welcomed the crash of waves.

This was a night created for ghostly fears. She strove to banish them by attending to familiar things. She looked at the sleeping children; set the table for Carlton's supper; laid out dry clothes for him; placed his worn old slippers on the jutting hearth of the stove. She tried to replenish the fire, but the moisture-laden air stifled the open drafts and caused wisps of acrid smoke to curl around the black covers and add to the dimness and suffocation of the kitchen. She went outside and sat on the top step of the porch where she was instantly veiled and swathed in clinging mist. She drew a cigarette from the pack in her pocket; but the dampness of her fingers permeated it even before she had failed to light the match.

She thought of the dead woman in her own silent, fog-wrapped house and of the generations of men and women who had lived there, for one hundred years and longer, so people said. She saw Shag Island in its sodden desolation and the work which the three men had gone there to do. She saw the coming of the winter, this late summer fog its harbinger, this early darkness its forerunner.

When at eight o'clock she heard the muffled throbs of Carlton's engine, the dulled scrape of the dory on the beach as they hauled it higher, and the sound of their voices in the road below saying good-night, she had become weak and spent from tears.

4

As Carlton came with long strides up the path to the house and followed her inside, there was an odd restraint within his usual banter which she was quick to discern.

"No skyscrapers to see tonight behind our city," he said. "No floodlights on our Fifth Avenue. As for Shag Island, it's a safer place to be buried in than to live on, even for two hours, in this fog."

He was drenched, standing there in the middle of the dim kitchen, water trickling down his rubber boots in rivulets on the floor. She could smell the musty stench of wet wool, soaked with earth and salt water. He looked at her with questions in his eyes which he did not expect her or anyone else to answer. Then he suddenly gathered her in his arms, holding her close in a need of which she had never before been so conscious.

"Get those wet clothes off," she said, when she could speak. "There's dry things on the bed, or they were dry half an hour ago. I'll get your supper right away."

"Have you had yours?"

"No, I couldn't eat."

"Good," he said. "I don't hanker after eating even a steak, alone."

"You needn't worry about any steak," she said, as he let her go and went toward the bedroom. "You'll get a can of beans if I can heat them up at all on this pesky stove."

"Good," he said again.

He came out a few minutes later in his red flannel robe, which she had made for him years ago as a birthday gift. For the first time she could remember, his tall, strong body looked stooped and tired; and there were lines about his eyes which she had never seen there. When he had stowed his boots and oilskins in the shed off the kitchen and washed at the sink, he carefully moved her plate, her cup and sau-

cer, next his place at the table. Then he lifted her chair
and put it beside his own.

Vast waves of gratitude swept over her as they ate their
supper. She had never understood that sadness can unite
as well as separate, heal as well as hurt, that fears, once
recognized and shared, lose their trailing shadows. Now and
then he laid down his knife and fork to place his hand
upon hers, to follow the outline of her fingers with his own,
then to hold them tightly in his grasp. He buttered slices
of bread for her as he had foolishly done years ago, smiling
quietly as he gave them to her. She did not ask him what
had happened on Shag Island, nor did he tell her so much
as a word. When he spoke at all, he spoke of the weather
for tomorrow, the hopelessness of any change with the wind
as it was. She wondered why she was not crying. Perhaps
she had cried herself out; or just perhaps there were times
and things, as in some dimly recalled poem she had once
known, too deep for tears.

She wondered, too, where her fears had gone when an
hour later she lay safe and warm in his arms. It was then,
when she felt his body loosen in sleep, that she decided to
repaint his buoys for him, as a surprise, once this beastly
weather had changed.

Drusilla West

◆ ────────────────────────────

DRUSILLA WEST, known as Trudy by her closer associates, who thought her real name both absurd and ill-fitting, was faced by a problem on the morning of old Mrs. Holt's funeral. She decidedly did not want to go to the funeral, but she could not discover any way to get out of going. If she gave sickness as an excuse, her state of health was bound to be wrongly interpreted, even by such a kind woman as Lucy Norton. If she said that company was arriving at just the wrong time, her guests would be subject as always to suspicion. All the few arrivals in this miserable place, whether clam-diggers driving through to Mackerel Bay, or duck hunters making for even lonelier coves and inlets, or unidentified strangers of any sort whether by land or by sea, were scrutinized and labeled by everybody within half an hour. Also, to add to her dilemma, there was the added embarrassment of her little boy, who was now out somewhere picking flowers.

When she had got up late that morning, she had found him in the kitchen eating some cornflakes at the table.

"What do you think you're doing?" she said crossly.

"Just eatin' my breakfast," he said, looking at his mother with a curiously self-assured expression in his round blue eyes. "We're startin' out soon."

His stare made her uncomfortable and irritated.

"Starting out?" she asked him. "Wherever to?"

"I wouldn't know just where," he said slowly and with an annoying precision, "but somewhere after flowers for old Mis' Holt. When we've got a great number, we're carryin' them over to her house. We aim to cover the whole scow with them."

"Scow—whatever do you mean?"

The child gazed at her patiently.

"I told you yesterday afternoon," he said, "but you wasn't listenin' with all those folks around. I told you that they're takin' her over to Shag Island to bury her. She lived there once, a long time ago. They're takin' her on a scow behind a motorboat just after the funeral. An' all of us children are goin' to cover the scow with flowers."

"That's what *you* think," she said. "You're not going to any funeral. Children don't go to funerals."

He quite calmly took his bowl and spoon to the sink where he rinsed both carefully with water, which he dipped from the zinc pail on the shelf, and dried them on a dish-towel, which he drew down from the rack. He had to stand on tiptoe to reach the towel. Then he carried the milk-pitcher and the sugar-bowl to the cupboard.

"We're all goin'," he said. He did not seem in any way to be disputing his mother's statement but instead only making another of his own. "Mis' Norton has it all planned out. We're sittin' on them little chairs in the kitchen where we used to read our books to old Mis' Holt."

He drew his sweater from the back of his chair by the table and drew it on over his blouse and his thin shoulders.

"I don't believe you've so much as washed your face this morning," his mother said. "You look a sight."

"I have," he said. "There wasn't no hot water, though, an' the wood was too wet to build up the fire. I tried, but I couldn't make it go. There ain't no kindlin' left."

"Say that again," his mother said. "You're beginning to talk like all the riff-raff on this coast."

"There isn't any kindlin'," he said, still patiently.

He walked across the kitchen to the outside door.

"Where are your rubbers? It's soaking wet everywhere after all this fog."

"They're out on the porch," he said quietly.

He opened the door and started to go out. Then he turned and faced his mother.

"You can give back that pistol to the one named Tony," he said without the least trace of emotion in his voice. "You can tell him thank you, but say that I don't have no use for it. It's in its box on the sittin'-room window sill."

"You're an ungrateful little pig," his mother said.

She moved quickly toward him, but stopped suddenly before his complete composure.

"Well, I'll be seein' you later," he said.

2

After he had left the house, she had after much work got a fire going, smashing up a clam-basket behind the stove in order to get some dry stuff which would take a match. Then she made some strong black coffee and returned to bed with a pot of it and a cup and saucer. Now she was sitting up in her night-dress, against the messy pillows, drinking the coffee slowly and smoking cigarettes as she reconsidered her problem. The fact that she had only a few cigarettes left added to her annoyance. She would have to go to the store after some more, and she did not relish that thought, once she had seen Hannah Stevens come down the road to take over for Lucy Norton. She was embarrassed also over the necessity of charging them.

She had hoped and planned on going into town to the factory this morning to pack sardines. If she had only been able to manage that, this awkward question of the funeral would never have arisen. There was every chance of work since herring were running in the bays eastward, as she had learned from one of her friends, who had come to see her last evening. He said there would be work for some weeks and, since she was more nimble at it than most of the women who packed, he felt sure she could put in all the hours she wanted.

Transportation was always the trouble with work at the factory. She couldn't be seen going into town with this friend if he lingered until dawn as he sometimes did. There was just too much activity going on at dawn among these wretched fishermen. One of her constant anxieties was getting her visitors out of the house before the fishermen were up and about; and her concern was in no way lessened by the fact that they were often in no condition to leave, or, for that matter, she, to drive them out. Joel Norton was always willing to take her in when he went in his truck; but even though she could weather the embarrassment and discomfort she always felt with Joel, friendly as he was, there was no sure guarantee that she could get a lift back at nightfall. She had lately been pinning her hopes on the Randalls, for Amy Randall now and again packed fish, and Jim was always on the road going somewhere or other in his car. But they had not told her that they were going in, and if they had gone earlier as might well be possible, they had either not troubled to stop for her, or she had slept too late and too heavily to hear them. She felt furious with them for neglecting her and then furious with herself, or perhaps more truly with a situation that reduced her to resorting to them.

She again returned to her problem. She simply could not bring herself to go to that funeral. In the first place, she had not known the old woman at all during the one endless year she had endured in this dreadful place. She had actually seen her only a few times, and then only when necessity had compelled her to fetch her little boy, who seemed to have a mania for running over there, either with the other children or by himself. Upon those few occasions she had felt distinctly uncomfortable, although she could not say exactly why. In the second place, she did not welcome the stares of her neighbors and the comments which she knew very well they always made about her and the small pleasures she enjoyed. After all, everyone to their taste, she said, flicking the ash from her last cigarette into her saucer. And in the third and last place, funerals made her jittery and depressed, and, as God well knew, there was quite enough without them to make her feel alone and hopeless.

3

With a curious inconsistency she dressed herself carefully and neatly after an equally careful wash at the kitchen sink. She looked fresh and pretty in a blue cotton frock and a pair of red shoes, which she stared at a bit ruefully as she slipped her slender feet into them, since they were not yet paid for. Then she covered her dress with an apron and thoroughly cleaned her small house. She went at this task with an almost feverish energy, scalding and polishing the sticky tumblers on the kitchen shelf, disposing of some empty bottles in a large paper shopping-bag, which she hid beneath the sink, putting fresh sheets on her bed and seeing that her son had made up his cot neatly in the sitting-room, sweeping and dusting thoroughly. In some odd way she always felt better when her house, at least, was put to rights; and when she had finished with it, it left little to be desired in cleanliness and order.

She wanted some cigarettes terribly, but she would do her nails, she decided, while she summoned up courage to run down the road to the store. Also over her nails she could compose in her mind a letter to her husband, which, since she saw no way of getting into town to the post-office, she would give to the mailman when he drove through at noon. In the other places where she had lived, she thought, as she carefully spread the polish and looked critically at her smooth, white hands, rural drivers would give one a lift now and then; but to expect any such favor from this doddering old man was out of the question. Besides, she was frightened of him and his nosy ways.

Once her hands were as she wanted them and she had brushed her short, fair hair before her bedroom mirror, arranging it simply and even demurely with a black velvet band around her head, she took off her apron, told herself she was an idiot to be afraid of Hannah Stevens, and went to the store to get her cigarettes. Hannah, in spite of herself and contrary to all her moral and religious standards, was impressed by her polite and cordial manners, her simple and tidy appearance, and her gracious comments upon how wonderful this change in weather was. Fortunately for them both, the funeral was not mentioned.

4

When she got back home, she sat down on her porch steps for some long and satisfying draws on two cigarettes and for some more careful planning of her letter before she went indoors and sat down at her sitting-room table. From its drawer she took a box of pale blue stationery, heavily engraved at the top of each sheet with the form of her name current among her friends. After a few moments of consideration she substituted for it some humbler sheets from a cheap tablet.

"Dear George:" she wrote.

Then, because she felt lonely and uncertain and did not know quite what to say, because the dread of old Mrs. Holt's funeral weighed upon her, and because the vast reaches of the sea made her nervous and sad, even in this new sunlight, she went to her bedroom and cautiously measured out some gin from a bottle which she drew from a drawer of her bureau. She hesitated for some moments before she poured it into the tumbler, and she was very careful about the small amount which she allowed herself. Then she took the tumbler back to the sitting-room, lighted another cigarette, and between small sips of the gin, took another sheet from the tablet and began her letter again.

George:

I am writing to you again because you have not answered the letter I wrote to you three weeks ago. I told you then that I needed money, and I need it even more now. And you would better keep your dirty mouth shut about my earning any on my own.

What I earn, I earn from packing sardines in the factory and getting an awful backache doing it. But I'd rather die packing fish than ask you for a cent, if there was any regular way to get out of this place to the factory, only there isn't. You might just remember that I need the money most *for your son*. He *is* yours, you know that as well as I do, or you would if you lived here a while and saw how much he is like you. He's getting ruined in this hole. He won't mind me. He does just as he pleases, and he gives me the creeps the way he stares at me and then acts as he likes. If your mother wants him, she can have him. She'll get a prize package, believe me.

You may like this house where you was born, but I hate it. That foolish couple up the road named Sawyer would like to buy it. They say it's better than their place. Ha, Ha, I say to that. One of these shacks is no better than any other in this dump where you dumped me and that stinks more every day.

Now you just get out the old pen and tell me what you aim to do about the boy and me. I've had all I can take of this God-awful lonesome place where there's nothing but wind and fog and the roar of these waves. And you get yourself to some post-office and send me off a money order for a good fat sum. I can't go on running up bills at this store. I've got my pride if you haven't. And it don't help me one mite to have those Nortons telling me what a nice little boy you was once. *Once* is right. They're telling me!

I'm warning you now for the teenth time that I'm clearing out. I can get me a job any day in a lot of places where I won't go dotty as I'm going here. So you just stop throwing cash around on you know what and send some to your wife and child.

D. W.

P.S. That old Mrs. Holt you used to like when you was a kid in this hole is dead. She was 90 years old. I'm 62 years younger than her, but there's lots of times when I wish I was dead, too. And, believe me, this is no joke either. So just don't take it as one.

5

She felt weak and sick when she had finished her letter, addressed it, and rummaged around until she had found a stamp for it. She would have liked a spot more gin, but she resolutely kept away from the bureau drawer. She knew with a real and honest dread that she was becoming far too dependent on the lift it gave her out of her depressions and despair. She knew, too, that after the lift was over, the despair which followed was beyond all description. And she was not at all unmindful that gin was one thing that you couldn't charge, and that unless her friends chose to give her some or to leave a bit after their visits, she would be hopeless all the time instead of almost always.

She went into the kitchen, cut some bread from the loaf she had brought from the store, and made a sandwich from a can of deviled ham. She heated up the coffee in the pot, using the remnants of the clam-basket to start up the dead fire. Her letter was in her apron pocket. Once she heard the rattling old mail truck coming over the road, she would run out and give it to the driver, who rarely drew up at her lopsided aluminum box on its crooked post.

She was still trying to extricate herself from the net in which the funeral had caught her when she heard the truck careening through the ruts and gravel of the road. Hastily wiping her mouth and hands, throwing off her apron, and giving an instinctive touch to the ribbon around her hair, she ran out to meet it. Hope and relief quickened her heart when she saw, swaying down the long headland hill and approaching the mailbox where she stood, another truck loaded with clam-digging gear and with three men on the single seat. The clam-truck drew up in the tangled grass on the far side of the road to wait for the mailman to drive on.

"Traffic," said one of the men. "First time I've ever stopped for traffic on this boulevard."

Two of them looked appreciatively at her as she handed her letter to the silent mailman. When he had gone on, they still lingered.

"Well, what do you know?" the younger of them said. "Ain't I seen you somewheres before?"

"Most likely," she answered, with a captivating lift of her head. "I get around."

"So I'd guess," the other said.

The older man, who was driving the truck, said nothing and began to shift his gears.

She seized upon this moment of salvation for the immediate present, not to say of possible compensation for the future.

"I don't suppose you could take me through to the main road," she said. "I've got work at the factory, but my friends didn't know I was packing today and went off without me." She shot an arch glance at the youngest clam-digger. "There seems to be plenty of room in the back," she added, to him.

He gave her an appraising smile.

"There's plenty of room right up here in the front," he said. "Get your trappin's and come along."

She started toward the house. For an awful moment she thought she was going to burst into tears from relief and excitement.

"How do you plan to get home?" the driver called to her. "We don't often make low tides way out here."

"I'll take my chances," she called back to him.

The two younger men laughed and nudged each other.

"I guess she won't have to worry too much about gettin' back tonight," one of them said.

"Or any night, if you ask me," said the other. "Just remember, will you, that I was the first to say we'd make room for her."

Daniel Thurston

◆ ————————————————————

AFTER THE doctor left Lucy Norton at the store on the night of Sarah Holt's death, he went on up the long hill through the fog and rain to see Daniel Thurston. It was past midnight and far too late for calls on anyone at all; but since someone had telephoned his office that afternoon that the old man wanted to see him, he would take the chance, especially as he had no intention of making a journey like this one on two succeeding days.

As he shifted his gears among the jutting rocks of the hill and cursed his lights which made almost no impression against the fog, he wondered again what he was doing for his sixth year on this outlying coast. Two or three times during every one of them offers of practices elsewhere might well have tempted him if he had had any leisure to consider them, good openings in civilized towns where patients presumably not only paid their bills, but could tell you what ailed them in recognizable language rather than in taciturn growls or stubborn evasions. In such places he would have the chance to attend clinics, confer with specialists, and diagnose other illnesses than rickety hearts caused by hauling heavy lobster traps and pulling oars in rough seas, stomach ulcers from wrong food and too much of it, and arthritis, that last haven toward which most fishermen traveled by way of cold and exposure and fogs like this one. Just why at this hour of the night he was going to see an old scoundrel, who should rightly have died of a worn-out heart and high blood pressure three years ago when the doctor had first seen him, he was quite unable to say.

Once he reached the top of this hill, if he ever did, he knew he must turn off to the left and bump for a quarter of a mile farther along a rough wood-road, which led sharply downward through dense undergrowth to the house where Daniel Thurston lived. He was startled when, halfway up the hill, he saw through the fog the intermittent beams of a flashlight near at hand and recognized behind it the figure of a man standing by the side of the road. The man dimmed the light quickly, but not before the doctor had jammed on his brakes.

"That you, Randall?" he asked sharply. "A bit late, isn't it, for you to be up and around?"

The man hesitated, but only for a moment.

"Fact is," he said, "I thought you might be along. I knew old Dan wanted to see you."

"So you were waiting for me," the doctor said, with more than a tinge of irony in his voice, which was not lost upon his unexpected companion. "What's the matter with Dan?"

"I don't think it's him," the man said. "I think it's his dog."

"His dog! Well, I'm damned! I've already had a night of it without seeing to a dog."

The man again turned on his light, holding it so that the outline of the doctor's car was visible in the fog.

"Dan sets store by that dog," he said. "The dog's all he's got, you might say. He thinks it's goin' to die."

"Why he isn't dead himself, I wouldn't know. Who looks after him anyhow?"

"Well, he does pretty much for himself. My wife takes down a bit of cookin' now and then. And my little girl goes down quite often after he's in from haulin'. They play school."

"They—what?"

"Play school," the man said. He had an oddly gentle voice, which at the moment puzzled the doctor. "Dan somehow got one of them old-time benches an' a desk out of that schoolhouse at Mackerel Bay before they tore it down."

"I'll bet he did—*somehow*," the doctor said.

"Well, I don't s'pose them old benches and desks amount to much. Dan used to sit in the one he got, when he was a boy. He's kind to children, an' he's quite set on my little girl. They like playin' school."

"What time is it?" the doctor asked, holding his wrist to the dashboard. "This fog seems to have stopped my watch."

"It's goin' on for one o'clock," the man said quickly.

The doctor released his brakes, and the car slid backward in the gravel.

"I hear how old Mrs. Holt is failin' fast," the man said.

"She's dead," the doctor said, as he crunched forward.

"Take it easy now in them ruts on that path," the man called after him.

2

The doctor reached the top of the hill, turned to the left and swayed, still in low gear, for a quarter of a mile along the rough road which led sharply downward through the drenched alders. Now he felt the wind from the sea, cold and fresh. The fog was somewhat less dense on this headland, in the shelter of which, in a cleared space above the shale and boulders of the shoreline, stood a small red house, set about by neat piles of lobster traps and flanked on the left by a low gray shed for the storage of fishing-gear. He hurtled toward the clearing, brought his car to a stop before the house, turned off his lights, and moved toward the two front windows.

He could see clearly into its one room, which was lighted by a large kerosene lamp on a square table in the center, covered with a white oilcloth. It was a room of considerable size, scrupulously neat and tidy. There was a black wood-stove in the corner to the right of the one door. On the wall next the stove were shelves for dishes, all in order, and hooks for cooking utensils. Along the same wall was a sink with a green iron pump attached to a drain-board. On the opposite side of the room there stood against the farther wall a school-bench and desk, and on the wall itself was tacked a large square of tarred paper which clearly served as a blackboard. In the corner to the left of the door was a bed of the old-fashioned spool variety, now disarranged by blankets piled in a heap upon it and surrounded by pillows.

By the side of the bed an old man was standing and slowly waving a palm-leaf fan. He had not seen one of those fans, the doctor thought as he stood looking through the window, since he was a boy and went to church with his grandmother in the Maine town of his birth and upbringing. When he started to move toward the door through the wet, unkempt grass, he saw the old man leave the bed and come to open it for him. Then he saw that a dog was stretched out on the heap of blankets, his head carefully raised upon one of the pillows.

The old man opened the door with no greeting, but with evident relief. He was tall and stooped in the shoulders. He wore a blue flannel shirt, open at the neck, and shapeless brown corduroy trousers. His thick white hair was neatly parted and brushed, and his heavy white eyebrows shadowed eyes common to those who follow the sea, quick and wary, but with a slightly glazed expression. His face was far too red, the doctor thought, but that could be explained by the intolerable heat of the room. His teeth behind his parted, cracked lips were in need of attention, two gold ones alone holding firm among the tobacco-stained stubs.

The doctor spoke first.

"What in hell's the matter here, Dan," he said, "that you rout me out on a night like this?"

The old man closed his mouth tightly, moved toward the bed, and, picking up the fan, began to wave it to and fro above a shaggy red dog of no discernible breed which lay quietly on the heap of blankets. The doctor looked closely at the dog.

"What's the matter with him?" he asked.

"His nose has been runnin' for two days, and he wheezes awful bad. He can't breathe right unless I fan him."

"How long have you been at that?"

"Hours, I guess."

The doctor took off his coat.

"It's too hot in here for both of you. Open that door or a window."

"I ain't openin' nothin'," the old man said. "He shivers too much."

"What have you been feeding him?"

"A bit of hot milk. But he don't fancy nothin'."

The doctor bent over the bed and examined the dog while the old man still waved his fan. Then he straightened up and took him by the arm.

"You sit down here," he said, leading him to a rocking-chair by the table. "Just sit here till I get some hot water for this brandy. I'll tend to your dog."

He came back from the stove with a steaming cup.

"You get this inside of you," he said, "and leave that fan alone."

The old man sat obediently in the chair and sipped the hot drink. He did not want to sit there, but for some time he had felt with fury a swaying in his knees and feet, and he knew that he was beaten. He moved the rocking-chair so that he could watch the bed.

"That dog don't take kindly to strangers," he said. "He's only used to me."

In a few minutes the doctor came back from the bed. He drew up a straight-backed wooden chair opposite the rocker and sat down.

"The dog's dead, Dan," he said quietly. "He wasn't so much sick as just old and worn-out. How old was he?"

Dan did not answer for some time. The doctor, watching him, saw that his knees were shaking inside the corduroy and that he could hardly raise the cup of brandy to his mouth. The doctor opened the door. The wind from the sea blew in, fluttering the oilcloth table-cover and the flame of the lamp, raising the long red hair on the dog. The old man got up, walked unsteadily to the door, and closed it. Then he sat down again.

"Thirteen," he said.

"That's old age for a dog," the doctor said. "That would mean more than ninety years old in a human. You'd best get you a new dog. There's a fellow over my way who's got some nice pups he wants to give away. I'll bring you out a good one when I come again in a day or two."

"I don't want no other dog."

"All right. Suit yourself. But there's one sure thing, you can't stay in this hot place with him here. Let that blasted fire die down, and get some sleep. You can't haul traps today or tomorrow either. I'm leaving you some more brandy and these pills here. You take them both and get to bed. I'll wrap your dog up and take him along with me. I'll bury him in my rose garden. He'll like it there."

"Don't you touch my dog," the old man said fiercely. "He ain't layin' in no rose garden. I'll find a place for him when I go out in my boat soon as it's light. You leave him be."

They sat silently in the hot, close room, the sharp, metallic ticking of an alarm clock on the shelf above the sink cutting off the minutes.

"Well, if you won't lie down," the doctor said at last, "and you won't let me touch your dog, you can at least open up your shirt and let me see how you're making out."

The old man did not move his hands.

"I ain't openin' no shirt," he said. "I didn't get you out here to fuss around me. You get goin' on your homeward way. I'll see to things here."

"I'm not moving a step until you promise me you'll haul no traps today or tomorrow. You beat all the beatingest fellows on this coast. Hustle up now and get that promise out of you, or here I stay."

"I won't haul. I wasn't reckonin' on haulin' today anyhow. The fog's too thick."

"Okay, then. What about that dog now? He can't stay here."

"You leave him be!" the old man said.

3

He opened the door for the doctor and stood in the glow of the lamplight until the car had turned in the clearing and disappeared among the alders. Then he shut the door, shifted the dampers of the stove, and half closed the flue of the rusted stove pipe. He walked to the bed. Raising the stiffening body of the dog from the pile of blankets, he moved it carefully, but with seeming unconcern well over to one side, heaped the pillows against the headboard, and, blowing out the lamp, lay down on the outside of the bed, covering himself with one of the blankets. He lay there quietly although he breathed with difficulty, until the dark windows of the house turned pale with early light.

The fog was still heavy when he opened the door, but the rain had ceased. Until he smelled the dank odor of the mud and flats, he had forgotten that the tide would be out. He could hear the muted engines of his neighbors getting underway for the morning's haul. He waited until all was quiet in the harbor. Then he crossed the wet clearing, opened the unlocked door of the shed, went inside, and drew a square of canvas from among a pile of spruce withes and laths. He spread the canvas carefully on the floor. When he returned to the house, he left the door open, went to the bed, and lifted the body of the dog. It lay straight and rigid in his arms, its legs extended. He carried it outside and into the shed, which he locked carefully half an hour later.

By the time he had spread up his bed and put more wood in the stove, it was six o'clock. He got ready for the day as he had always done, although he angrily discovered that he was slow and awkward about the most familiar things. When he shaved at the sink, he had to brace his body hard against it; when he looked into the mirror, he could not see too clearly; when he went to the chest of drawers against the back wall of the room to get a clean shirt, he found himself suddenly in his swaying rocker. A sharp pain in his stomach made him give up the notion of breakfast; but he made some coffee and brought a slopped cup and saucer unsteadily to the table. As the pain grew worse, he managed to pour out some of the doctor's brandy and to swallow it with the pills which he had left. Then, furious with himself and with things as they were, he lay down

on his bed. Almost at once he felt warm and comfortable
and forgot his anger in sleep.

4

When he awoke, he was startled to see that his alarm
clock said nearly four o'clock. While he had slept, a whole
day had passed, and, of far greater importance to him, a
tide had come in and was now on its way out. This frightening knowledge made him hurry outside as fast as he
could to scan the still fog-wrapped shoreline. His dory
was afloat; but by the time he had accomplished what he
had planned and returned home to his strip of beach, he
would have to manage a tough walk over mud and flats and
slippery rockweed. With such an inevitable sequel, he felt
too dazed and tired to undertake his task although the ad-
mission of the fact exasperated and angered him. He felt
carefully in the pocket of his shirt to be sure that the key
to his shed was still there. Then he returned to the house
and began to think unwillingly of the necessity for food
and coffee.

His fire had become dead ashes during his long sleep. The
chore of rebuilding it seemed impossible to him. He sat
down for a few minutes in the rocking-chair by the table.
The late afternoon was coming on, and the room grew
steadily more dim from the thickening of the fog outside.
He thought he would best light his lamp; but that, too,
seemed to demand a strength which he did not possess. The
days are surely drawing in, he thought, staring through
his windows at the misty outline of the dark spruces. After
a little while he crossed the room to his bed and lay down
again.

(155)

The Randalls

✧ ─────────────────

WHEN THE doctor had gone up the hill, Jim Randall returned quickly through the fog to the ramshackle porch of his dark house. He found his wife, Amy, waiting for him there. She was sitting on a broken-down old stool, and he could see by the flash of his light that she was trembling with fright. He was conscious of an almost acrid sense of distaste for her, and then felt suddenly sick, and ashamed of himself.

"My God!" she said, almost in a whisper. "What you goin' to do now?"

"The only thing I can do. I can't take no chances on him comin' up from Dan's an' findin' us all here. He's no fool. Those fellows are late. I've got to meet them somewheres along the road, even if 'tis out of the fryin'-pan into the fire. Then I'll turn off somewheres else an' wait till he goes by. Step lively now an' help me get them things on their way."

She followed him in the wet darkness to a pile of brush behind the house. In the uncertain gleam of his light she helped him uncover the dressed carcasses of two deer and carry them to the back seat of his car, which was drawn up at the side of the house. They were heavy and awkward to handle and almost impossible to conceal once they had been crammed into the small space, though they did their best to cover them with pieces of canvas.

"Didn't Dan know?" she asked, still in a half whisper, as they worked frantically to stow away their freight.

"I told him, careful as I could," her husband said, "but he's likely to forget. He's gettin' old, an' that dog was all he had on his mind. He's not what he used to be in the rum-runnin' days. He really was somebody then."

"I was scared out o' my wits when I heard you talkin' to that doctor," she said.

"Who'd have looked for him this time o' night? I didn't get along with him too well either. He's a sharp one. He sees an awful lot around these parts. He's just lately been stirrin' up a lot of trouble with them game wardens to the eastward. He needs watchin'."

"You hadn't ought to have listened to Dan two years ago," she said.

Again he felt creeping over and inside him a loathing for her, trying her best to help him.

"Shut up!" he said angrily. "There's a lot o' things I hadn't ought to have done. It's a pretty time now to cry over 'em! Go to bed. Don't look for me till I get back, an' don't have any lights showin' around here."

He got into the car, started it almost noiselessly, and drove out of the yard.

2

After she had heard him easing his way down the hill and finally lost all sounds except the dripping of the fog from the trees and the intermittent surges of the waves against the headland rocks below, she went indoors and groped about in the darkness until she had drawn some thick, make-shift curtains across the two windows of the one-room shack. When she had seen that the door to a sort of ell in which her little girl slept was tightly closed, she lighted a candle, heated some coffee in a battered coffee-pot over a can of Sterno, and sat down by the one table in the dim, damp room to drink it. She wished almost desperately that she had some whiskey to make her forgetful and sleepy. Whiskey was hard to come by, and her husband never touched it. He needed all his wits, in his job, he said.

She asked herself while she drank the bitter, warmed-up coffee just why she was still fond of him, for he, she was quick enough to see, had often not much love for her. He was just off and on, now here, now there. You just never knew how he might take a fancy to be. She had been a fool to be taken in by him at the start.

He had told her when she decided to marry him ten years ago that he was a mechanic by trade. He could repair any engine that ever ran, he said. He warned her that they might rove about a bit, looking for work in the best places, preferably out-of-the-way ones, where there were plenty of engines, but few mechanics to deal with them. At that time, sick to death of the chain-store where she worked and of querulous, fussy parents who disapproved of all that she did, to rove anywhere at all seemed more agreeable than anything she had ever known.

When she discovered that roving usually meant a narrow escape from too acute game wardens and once from a threatened sojourn in jail, she was more disturbed and nervous than morally outraged. She was not at all prudish so far as morals were concerned. She just felt that life was an unfair game, in which some folks were lucky and others unlucky, and that, if you could beat it on any terms, so what? In fact, it had been rather a lucky day for Jim Randall when, wandering through the chain-store to take an account of how things were run there and to determine whether shoppers were actually what they seemed to be, he had found her, pretty, if sharp-faced and designing, behind a counter.

She held her wrist-watch close to the candle flame. It was a nice watch, a Swiss one, her husband said, when he had given it to her at Christmas a year ago. He had gotten it in a lottery on one of his frequent trips to some place or other which he rarely specified. She was proud of it since it compensated for her few clothes which she was always making over for herself or cutting down for her little girl. It was already long past one o'clock. She would make some more coffee, she decided. It couldn't make her any more jumpy than she was.

She had just measured out the coffee for the pot, poured in some water, and was getting ready to light the can of Sterno from the candle flame when she heard the sound of a car. Her heart leaped for a moment until she realized that it was not coming up the hill, but down. She blew out the candle and made her way to the front window, cautious against drawing the curtain so much as a crack until the car had passed, so that its lights could pick up nothing even if they could pierce the fog. Now she tried to recall the possible turn-offs on the long road and to imagine, if things had gone as they should, in just what thicket her husband was waiting until the doctor's car had passed and he was safe to come home.

As she drank the fresh, hot coffee and, to her vast pleasure, unearthed a few cigarettes in a crumpled package from the pocket of an old jacket hanging on the wall, she returned to the question of what there was about Jim Randall which still made her stick to him. She felt sure it had nothing to do with the fact that neither of them had looked elsewhere after other men or women, though she prided herself more than a little upon their mutual faithfulness, or at least upon hers to him. She guessed, perhaps, that it was because of his kindness to her when now and again he would take a steady job. She knew better than to urge these upon him. He just unaccountably decided to change his precarious ways and to settle down for a few months at steady employment and regular wages.

He was an excellent craftsman. There was literally nothing he could not do with his hands whether at painting or carpentry. He loved making things and would spend hours working at a chest of drawers for his little girl or framing miniature furniture. Once in a sizable town he had taken a job with a dealer in antiques and within no time at all had built up a reputation for himself as an expert in repairing and refinishing. He had a feeling for tools and always kept the few he owned in perfect condition, a passion in direct contrast to any pride or care which he took in the tottering, tumbledown houses where they usually lived. Several of these he had built himself, never apparently wanting them to be anything but hastily improvised shelters for the time being.

Twice in their ten years here and there he had been the engineer on a summer cabin cruiser, making good money and coming home every fortnight for a few hours. He was not only good-looking in his blue uniform, but he occasionally brought home, together with the money, some clothes given him by the wife of the owner of the cruiser, when she had discovered that he had a family. Two or three times during the winter he had gone into the woods for some paper company to cut spruce into pulp lengths. And although she hated her barren existence in some place like this one where her neighbors were offish and suspicious, she was relieved by steady wages, the absence of fear, and, above all, by her husband's new ways. At such times he was genial and generous to her whenever he returned home, taking her to a show once in a while and sometimes even to a dance. At such times, too, he went out of his way to help the people whom they always lived among as strangers, shoveling snow for one, repairing an engine for another, seemingly taking pride and pleasure in his new status.

The status itself meant nothing to her, she thought, lighting one cigarette from another, except so far as their little girl was concerned. She did not herself care a nickel about what her neighbors anywhere thought of her housekeeping or of her. She had had far too many neighbors in her ten years with Jim Randall, all nosy, all suspicious, all trying to find out what with good luck they would never find out. But she did sometimes worry about the child, who had lately brought home tales of taunts and slights on the school-bus and from school itself, who was growing pretty and ought to have nicer clothes and a good time now and then like other kids. If she had had her wits about her ten years ago, or if her husband had not been such a silly fool when they knew the child was on the way, there would not be any such problem now. But there she was, and getting wiser every day about things in general.

She smoked the last cigarette. A frenzied search with the aid of a flashlight around likely places in the disheveled room failed to discover another. The coffee had made her so jittery that she was like to hit her head against the ceiling. Her Swiss watch now said nearly three o'clock. She blew out the candle and opened the door. This blasted fog, she thought. Yet she knew, as she cursed it, that their pursuits could hardly be carried on without its helpful concealment of shapes and even of sounds. She ought to go to bed, especially since with little money coming in and most of that owed, she would likely have to pack sardines for the next few weeks; but what with the coffee and being scared out of her wits besides, she couldn't possibly sleep. Then, just when she thought she really couldn't take any more shivers and shakes, she heard the unmistakable beat of their old engine coming up the hill.

3

Jim Randall had seen the car he was looking for coming along at quite too dangerous a clip three or four miles from the turn onto the main highway. He was at the moment out of his own car, which he had left in a convenient shelter a quarter of a mile behind. He stood by the roadside flashing some pre-arranged signals. If there proved to be a mistake about the car, on a road where driving at this time of night was almost unknown, he had half a dozen ways of getting out of that minor dilemma. There was no mistake. In a moment he was on the running-board of what looked to be a grocer's delivery wagon and proceeding toward his own hiding-place.

"You're late," he said to the two men on the driver's seat.

"Tire trouble," one said, "and a state cop besides. He didn't like either our time or our space. We had to lay off a bit until he had left our beat."

"I've had my own troubles," Jim Randall said, recounting these briefly. "You got to get underway in a hurry. Slow down here."

"How many you got?" the other man asked.

"Two good ones. I'm in here, up this cutting. Watch out for the ditch."

They made their transfer with no unnecessary talk. The bills which the driver of the long car passed to Jim Randall felt crisp and good in his fingers.

"A week from tonight," the driver said, "unless you telephone Jerry here. If you haven't got anything, just say, *I'll be seeing you at eight.* Nothing else, mind. If we don't get the message, we'll be in your yard around two o'clock. Don't go outside it looking for trouble. This is a risky game we're playing."

"You're tellin' me!" Jim Randall said.

When the long car had gone, he got into his own which he backed still farther into the uneven path among the undergrowth. Then he sat quietly in the darkness, his ears sharpened for the sound he wanted to hear. He was in no hurry since at last things had turned out again as they should. The money in his pocket answered all his minor problems. As for his major ones, he had not for years, if ever, been able to look squarely at them.

This latter fact sometimes troubled him, or, perhaps more truly, was simply a puzzle to him. It struck him as queer that a man could not account either for his actions or his feelings. Some odd impulse seemed to dictate his ways and means of existence. He did not particularly relish the hazards which he risked when he startled deer by the headlights of his car at midnight on some almost inaccessible back road, or took a long chance in tramping through the fog and rain in search of them in the woods; or when in the darkness he tampered with the padlock on someone's lobster-car. Perhaps there had once been excitement in these perilous undertakings, but it was not easy to recall any.

What was more perplexing to him was that he could not seem to summon up any regret either for his way of life or, worse still, for his behavior toward those persons who from time to time had trusted him and whom he had failed. His sudden decisions every now and then to go straight for a season and work respectably like other men did not seem to have their source in shame or in remorse. They also sprang from an incomprehensible impulse; and the pride and pleasure which he always felt from honest labor were apparently not sufficient to keep him from sliding back into his former habits and pursuits. In short, his life from his boyhood until now, when at the age of thirty-five he was sitting here in this dripping darkness, was like some riddle which children puzzle over for a few bewildered minutes and then abandon with careless disgust.

A small owl screeched from somewhere near at hand; and he wondered what prey it was finding in these wet woods. He was genuinely fond of wild creatures and would steal time even on some slippery errand to study their ways. He was always bringing home some small animal which he had trapped, or caught in his quick hands, to his little girl, a baby raccoon or a fox, a rabbit, or some squirrels or chipmunks. He liked taming these, teaching them to have confidence in him; he liked even more the companionship which they afforded him and the child and which often seemed the one bond between him and her. He could not reconcile his ruthless killing of deer at one moment and the pity which he felt at the next over the sight of an injured bird. Once, even while he was cleaning and loading his rifle in preparation for one of his illegal forays, he had grieved deeply and sincerely over the death of a ruby-throated humming-bird, which a high wind had dashed against a tree in front of his house. He had not been able for hours to get the tiny bird out of his mind. In short, he could not reconcile any of the divergent, contradictory traits which made up himself, nor was he seriously trying to do so as he sat waiting for the doctor's car.

If, indeed, he was focusing his errant mind on anything at all, it was on how soon they would best be clearing out of their present place of sojourn and where they would best go. Two years was a long time to linger in any one region, especially in one so circumscribed as this cove settlement and with but a single means of hurried exit, since he had no boat. He was perfectly aware that his doings were held generally in suspicion; but he was also comfortably certain that most fishermen kept their mouths shut, not because they lacked indignation, but because they feared revenge. He had the notion that he might try an inland area next, provided he could make the right connections. He had spotted two or three fields of profitable activity during his work for the paper company. The sea afforded too many approaches, and after a time it had a way also of getting on one's nerves. Take tonight, for example. If the tide had not turned two hours ago and the wind died a bit with it, this especial job would have assumed a far different complexion and taken a lot more out of him.

When he heard at last the sound of an engine throbbing dully through the laden air and caught the crunch and sway of tires in the gravel even before he saw the dim glow of lights, he smiled, not at the way in which he had outwitted the doctor, but at the singular fashion in which in moments of excitement the skin on the back of his neck tightened and the hairs rose, both together giving him a tingling sensation there. This had happened to him ever since he was a boy; and it never failed to amuse him and to stir his curiosity. It was oddly akin, though vastly more amusing, to the sharp, unpleasant quivering in his nostrils whenever he became unreasonably irritated or angry with his wife.

He sat motionless until, with the passing of the doctor's car, the tingling sensation passed also. Then, almost nonchalantly, he struck a match between his thumbnail and forefinger and took some draws on a cigarette. He was lingering in the hope of hearing the screech-owl again.

Since the woods were silent and he was beginning to feel sleepy, he started his engine, bumped carefully across the roadside ditch, and set forth for home.

4

His wife was waiting up for him. This knowledge, as he opened the door, filled his nostrils with the curious, distasteful pulsation, almost like an offensive smell. He did not reprove her for lighting the lamp, since early rising in a fishing community evoked no comment even if there had been anyone near enough at hand to notice it. She was starting a fire in the cook-stove to get him some breakfast; and the queer pulsation vanished as he saw how thin she was, how tired and scared she looked.

"I'm afraid you took an awful beatin'," he said, almost kindly.

She felt the puzzling attraction which he held for her coming back to warm the chill of the untidy room like the fire beginning to crackle in the stove.

"What about yourself?" she asked. "I'd say that was a close shave."

"I've had closer," he said, hanging up his wet coat and drawing off his boots. "Fry the eggs, will you? Is there any ketchup?"

"Yes," she said, reaching for the frying-pan with one hand and the messy ketchup bottle with the other. "Was everything okay?"

"Sure," he said. "Fifty bucks stowed away. It should have been seventy-five with the prices those New York fellows are ready to pay for venison; but I didn't take no time to argue."

The Backwater People

THE LONG, narrow neck of land which extended twenty miles southward and seaward from the main highway and upon which, two miles from its rugged, surfswept promontory the cove settlement lay, afforded subsistence also to other fishermen and their families. It was cut deeply on both its eastern and westward shores by the tides into any number of bays and inlets which on less rugged portions of the coastline would have been of considerable size, but which here seemed small and protected in comparison with the outer fishing-grounds like those facing the cove. Those who sunk their traps or laid their weirs in these inshore waters were known as "backwater folks"; and there was implicit in this term something more than a mere geographical distinction.

To the cove fishermen, those who fished the backwaters were a less adventuresome breed than themselves, less hardy and self-reliant, neither so enterprising nor so durable. Not possessing that spirit of resolution, even of defiance, which forms an integral and indispensable part of the off-shore fisherman, they did not stake their all on a gamble with open water where tides and winds and fogs were at their worst. They were content with shorter trap-lines which meant, to be sure, a smaller income, but involved less risk and a far smaller original outlay of cash. Nor did they depend entirely upon fishing as a livelihood. The fields about their widely scattered houses allowed them generous vegetable gardens, if they chose to plant them; they could keep a cow or two, and perhaps a few sheep on their rocky pastures. When fishing proved poor for a season, they turned to other forms of labor. They worked on road crews, or dug clams, helped to seine weirs in luckier bays, raked blueberries, or put in hours in the canning and packing factories of the towns eastward.

Certain of their activities were seldom wholly clear of suspicion. This uneasy feeling might well have survived from the fairly recent smuggling days, when the bays, inlets, and tidal streams which cut into their rough acres had been seemingly designed for the reception and concealment of contraband cargoes, and when their motorboats had regularly put out to sea under cover of darkness or of fog to meet some liquor-laden schooner from across the border. Now their talents in other illegal pursuits were likely to be put to use in nearer waters. Jim Randall felt far more at home among the backwater people than among his present closer neighbors; and Daniel Thurston had spent his earlier years among a former generation of them, although Daniel was an odd mixture of both cove and backwater. The cove fishermen, not without cause, thought their backwater neighbors a questionable lot and kept a wary eye upon them and upon their seafaring avocations.

2

The road which ran through their fields and pastures and past their swamps and estuaries had once been a thriving thoroughfare in the days when less had been demanded of roads. It had then connected the ship-yards at the head of the Tidal River with those at the cove and on Shag Island; and over it had passed a traffic which it would never know again and which was now, in its loneliness, almost impossible of conception. Oxen had hauled great loads of logs and timbers across its hard-pounded snow in winter. Huge wagons, drawn by three or four pair of horses, had carried furnishings and supplies for ships' cabins, slop-chests, and galleys. Shipwrights and sail-makers, riggers, caulkers, and painters had ridden over it on horseback at dawn and at dusk. Even coaches had swayed through its ruts, carrying gay people southward to launchings or northward to parties and weddings in coast towns, whose great white houses (now with welcoming Tourist signs in their windows and on their lawns) were almost the sole proof of quite another way of life. Over it, too, a century ago had passed seamen with dark faces and strange tongues bound for other forecastles on ships about to leave Shag Island for any one of a dozen foreign ports.

Its houses even today suggested more prosperous times. Of the twelve or so still standing along the twenty miles, each in its present state of dilapidation and decline was far superior, at least in its outward aspects, to any of those in

the cove settlement. Sagging sills and patched roofs, un-
painted clapboards and missing shutters, grassgrown drive-
ways and caved-in barns, could not entirely obliterate the
dignity they had once known. Two of them still retained,
somewhat back from the ragged roadside, iron staples set in
stone blocks, which had clearly once served as tethering-
places for horses. And before one of them there stood, even
yet, a great granite watering-trough, carved in a beautiful
raised design of leaves and flowers and still filled with running water from a nearby spring.

Few, if any, of the present backwater families were the
descendants of the original owners of these houses. The
owners now lay in untended family burying-grounds, and
their immediate descendants had long since gone elsewhere
after the coast had lost its shipping. Their houses had finally
passed into the hands of much less vigorous people who
came and went, bought for a song and sold for less, fished
and farmed desultorily for a few years, labored at this and
that, provided uneasy hours for shore commissioners and
for fish and game wardens, and then disappeared. Lucy and
Joel Norton during their thirty years in the store often
recalled this and that backwater jack-of-all-trades whose
accounts were still unpaid and whose present whereabouts
were unknown.

The shifting and shiftless ways of the backwater people
had long since lost them their school. Once this had stood
upon the hill above the inland reaches of Mackerel Bay,
seven miles from the cove. Now both backwater and cove
children were picked up by the school bus, which ventured fifteen miles along the road from the nearest town and

left any further transportation problem to be solved by the parents concerned. Joel Norton usually proved to be the solution.

3

In the summer months the backwater people made out well enough. Lucy, driving to town occasionally with Joel, envied them their sunny fields of daisies, buttercups, and orange hawkweed, their berry-filled pastures and tangled roadsides, their run-out apple orchards. They lived in a land of plenty, she thought. Often, when fog lay thickly over the cove, shutting out even Shag Island and the headland, their houses and gardens were flooded with sun. The fish drying on their rickety flakes were assured hours of heat and light. They did not have to compete with endless hours of dampness. Flounders swam up their inlets on every tide. In the spring smelts and alewives crammed their small brooks. They had land to till, which the cove people lacked, and places for hens and pigs. Their children had safe playgrounds. Even the helter-skelter piles of their few lobster traps and buoys held a reckless appeal, simply in their suggestion of an easier, less vigilant existence.

It was in the winter that the pinch came. Their inlets and estuaries, unlashed by the full force of the tide, were closed early by ice. The snow always lay deeper with them than the high winds from the open sea allowed it to lie at the cove. Their big old houses, all in disrepair, were cold and draughty. Whenever the doctor got through to the store for some hot coffee and a bit of shopping with the Nortons, he was always ready to talk about what he called "backwater complaints." He said he made it a practice to call at every house in addition to the one or two which had summoned him since he felt sure that someone was out of sorts in each.

Lucy owed her friendship with the doctor largely to the backwater people. When he had settled on the coast six years ago and begun to make his multifarious, far-flung calls, both by boat and by car, he discovered that Lucy was a mine of helpful information concerning the characters and backgrounds of those to whom he brought relief or rescue, or, perhaps, only companionship and courage. She knew the backwater, its unwelcome, sometimes unidentified babies, whom she had frequently helped to bring into the world, its old and infirm, whose departure from it she had often watched, its young people, to whom life seemed to offer little except monotony or short-lived, unwise fragments of sordid romance. She knew its assets and liabilities in spiritual as well as in material terms. The doctor had early fallen into the habit of consulting her concerning his patients and pensioners on the long neck of land, realizing that the information which he failed to extricate from them they had doubtless already divulged, or at least hinted to her.

4

When the winter really shut down over the backwater, it was to the store that its people came, except in the most impossible of weather. They drove through in their shabby cars and trucks or even returned with Joel from the town, and walked the long way home. Some of them, closer to the main highway, might more easily have gone to larger places of trade; but credit in these was not so generously extended or seemingly so unsuspected as it was with the Nortons. They always paid for their purchases if they could. Only after the most trying of summers and in the most ruthless of winters did they fall far behind. Aware of the attitude of their cove neighbors toward them as a whole, they were careful not to sink too low in whatever grudging respect they could command.

They were at their best in the store, enjoying its warmth and companionship, wishing their own houses were close enough together so that they might see friendly lights and run from one to another as the cove people could. When they bought far more essence of peppermint and vanilla than mere cooking could ever require and more Jamaica ginger than was necessary for a hundred upset digestions, Lucy quieted her sense of guilt just by looking at them and realizing what a toll their way of life exacted from them.

It was curious how within so small an area and so short a distance people could show such a contrast as that at once evident between cove and backwater. Backwater men slouched and ambled as they walked, were constantly shifting their positions as they sat in the store, were uneasy and uncertain in the way they looked at one. Their faces could take on quickly a churlish expression, whereas those of the cove fishermen were merely taciturn and withdrawn. They were given to disgruntled, ineffective protest, against the weather, the scarcity of lobster and herring, the cost of living. They quite clearly possessed few qualities of mind which could protest against or even recognize the far higher cost of life itself.

Few of the backwater women shared the standards of

housekeeping in which the cove women took such pride. Their curtains sagged at their front windows; their shades were torn and uneven; and their great old kitchens were likely to be both storage-places for fishing-gear and living quarters for their families as well as places for the hasty preparation of whatever food might be available. The clothes of the younger women were cheap and flimsy, though pathetically modish. Lucy was often uneasily certain that a new lipstick had been purchased at the cost of a sack of potatoes; and Joel procured unwillingly from the wholesale markets the cans and jars of fruit and jellies which, had they only been willing to salvage their fields and pastures in the berry season, they might have put up for themselves in preparation for the winter. Whether young or older, they seemed mere appendages to their husbands rather than sharers in any respectable way of life. They often looked tired and discouraged, and their too many children, scrawny, unkempt, and ill-nourished.

They were, all in all, both men and women alike, a thin-spirited lot. Lucy often wondered what really went on in their minds. But although, like all her nearer neighbors, she was given to looking upon several of them with caution, she would have missed them for far more than material reasons, had they not come frequently to the store.

5

Sarah Holt had known the backwater in far different days from these which Lucy knew. When she was young, its big houses had been well-kept and fine and had housed a quite different sort of men and women. From them had come first officers, young supercargos, and even masters of ships which had been built and launched in a dozen Eastern Maine harbors. In them had lived women who like herself had sailed to faraway countries and had known the world so well that they had never mistaken their native parishes for the whole of it, once they had returned to them.

"But you don't get anywhere by always looking back to the way things used to be," she said to Lucy. "It's the way things are now that we've got to live with. This coast once offered you the earth and on fairly easy terms, too. But now it's charging a heavy price for everything it has to give. And there's some people who just can't pay it."

That was true of the backwater people, Lucy thought, even as she sold them too much peppermint along with less exciting and dangerous provender. Most of them couldn't pay the price which their time and place demanded of them, and the reasons why they couldn't pay were quite unfathomable, at least to her. These lay buried deep within the obscure, unimportant history of every family who lived now in the old houses and, for that matter, along scores of other backwater roads which likewise had seen their better days.

When the news of Sarah Holt's death reached the backwater, several of the fishermen felt secretly an uneasy and even guilty sense of relief. Not many of them had known her well, but they had always been conscious of her presence and sensitive concerning her standards of human behavior. All the women were curious enough to call on their widely-separated neighbors or even to get through to the store in order to glean any added details concerning her funeral, toward which they looked forward in eager and pleasurable anticipation.

The Children

AT TEN o'clock the cove children gathered in front of the store to go after their flowers. There were six of them in all, none over ten years old. There were Hannah Stevens' two grandchildren, Marilyn and Benny, ten and seven, who had spent part of their summers with their grandparents ever since they were born and were inclined now to enjoy showing off their wider experiences to their country playmates. There were Marcy and Davy Sawyer, one nine, the other five years old. Marcy was a responsible little girl with red hair in two pigtails over her ears, blue eyes like her mother's, and a great concern for her brother, who was a frail little boy. There was Stephen West, who was ten and for whom the other children felt vaguely sorry because, even when he was playing, he had serious ways and always seemed somewhat anxious and old for his years. And lastly there was Elly Randall, whom they rarely played with for reasons not quite clear, but whom Lucy Norton had expressly named as the leader of their expedition. Elly had already assumed a position of importance in their eyes, partly because she clearly knew where all the best flowers were, largely because she was the possessor of ten cents, which at this moment she was spending in the store for candy for them all.

When Elly came out of the store with her sticks of licorice in a paper bag, they all started up the long hill. They were bound, Elly said, for a swamp well beyond old Daniel Thurston's house where she had seen some red lilies not two days before. All the children were excited at the thought of getting some unusual flowers for old Mrs. Holt, not just the common asters and goldenrod which were now blossoming in profusion along the roadsides, bent a little with the recent fog and rain and shining as the warm sunlight lay upon them.

"My grandma says there couldn't be any lilies now," Marilyn said. "She says they blossom in July and August, not now at all. It's much too late for them, she says."

Elly gave an embarrassed hitch to the straps of her faded blue overalls. She was somewhat in awe of the Stevens children, whom she had never played with; and even in this new importance of her position as leader, she felt uncomfortable and shy before Marilyn's words and beneath her unfriendly stare.

"I guess Elly knows," Marcy said quickly. "She knows more about flowers than the rest of us do. She wouldn't take us way into that swamp if she wasn't sure, would you, Elly?"

Elly felt that she must speak. Even the licorice was not enough to justify her at this moment. She turned her wide brown eyes on Marilyn's sharp gray ones.

"You can't never tell about flowers," she said quietly. "They do awful queer things sometimes. Those lilies stopped their blossoms in August, an' then a big clump of 'em began to put out some more. You'll see that I ain't tellin' lies."

"Your father tells lies," Benny said suddenly. "My grandpa says so."

Marcy seized him by the collar of his blouse and dragged him angrily away from the little group to the side of the road.

"If you don't want me to trounce you well," she said, giving him some premonitory shakes, "you stop sayin' those things to Elly. You don't belong here anyway, an' I won't ever play with you again if you say such hateful things."

She was relieved at Benny's sulky whimpering as she joined the others.

"That's true about flowers," she said to Elly, putting her arm around Elly's waist. "Once when my father took us on a picnic out to Hardtack Island, we found a blue iris in September. We took it home with us an' put it in a vase, but it was all crinkled up next day, not fresh at all like it was when it was growin' out there."

"That's the way with them irises," Elly said, her voice a trifle shaky. "Irises ain't pickin' flowers. They're more just to look at in their swamps."

She felt even more uneasy as they now passed her dooryard before taking the path through the woods past Daniel Thurston's house on the way to the swamp. It was cluttered with all manner of things, old tires, some broken crates, odds and ends of machinery. It had no bright flower-beds like the yards of the other children. Nor was her house neat and tidy like theirs. Had it not been for Davy, she would have led them all along the high path on the face of the headland, from which her house was not visible; but he was too little. There was one thing that comforted her, however. Her parents had gone away somewhere early that morning. When she had got up, they had already left.

"What does your father mean to do with all that junk?" Marilyn asked, scornful of Marcy's eyes. She was now holding her brother by the hand in case Marcy should spring upon him again.

"He can make something out of everything there," Elly said, with an attempt at pride. "He can make anything at all."

"An' he can fix things, too," Marcy said. "Once, before we had our new boat, he fixed my father's engine for him. My father said he never knew anyone who could fix engines so good as your father can, Elly."

Elly's brown eyes shone with pleasure. She suddenly loved Marcy with a strange, new love.

"I'm awful tired, Marcy," Davy said. "Couldn't we rest just a little mite?"

"Yes, we could, darlin'," Marcy said. She turned to Elly. "Let's us two make a seat for him with our hands so he won't have to walk down that rough road to Dan's."

They clasped their hands upon each other's wrists to make a chair.

"I'll lift him up," Stephen said, "an' if you get tired, I can carry him for a spell. He don't weigh nothin'."

"We'll rest, once we get past Dan's," Elly said. "There's a big ledge there lookin' out to sea. We'll eat our licorice there."

"It was nice of you to buy the licorice for us," Stephen said.

"Don't mention it," said Elly.

2

There was no sign of life around Daniel Thurston's red house which they skirted on their way to the ledge. The children saw that his boat was gone from his anchorage. Elly looked troubled.

"He wasn't supposed to haul today," she said. "The doctor said so. He's sick."

"I bet I know where he's gone," Stephen said. "I saw him edgin' round the shore just as you went in to buy the licorice while we was waitin' for you. He had a big sack o' canvas across his bow. He's likely gone out to bury his dog in the tide."

"He's a good, kind old man," Elly said suddenly, before anyone else could say what she was afraid of. "I don't know how he'll make out without Rover."

"I don't either," Marcy said. "He made Davy a boat once, didn't he, Davy?"

"Yes, he did," the little boy said. "It was a good boat, too. You can set me down now, Marcy."

They clustered together on the high, jutting ledge above the sea, all sobered at the thought of Rover being weighted down and dropped into the tide. Elly proceeded to lighten the atmosphere by dividing the licorice.

"What do you say we give two sticks to everyone but

us?" she whispered to Marcy, her face flushed with this new sense of friendship. "We can have one apiece."

"That's okay," Marcy whispered back. "That's just right."

"You don't have to give me two sticks," Marilyn said. "I've got my own pocket money. I can buy all I want."

"It's Elly's candy," Stephen said. "She can do what she likes with her own candy, can't she? I think she an' Marcy are very generous to us."

He took his two sticks gratefully and began at once to eat them.

"Thank you, Elly," he said.

The sun was warm and bright on the ledge. The sea stretched below and before them, wide and still. Marcy wiped Davy's mouth now and then with her handkerchief as he chewed his licorice. Across the cove they could see the Holt house in its high field. A man, clearly visible in the thin, transparent air, was walking to and fro along the beach.

"That's Thaddeus," Marilyn said, beginning to eat her second stick of licorice. "Don't he drink something awful?" She carefully put her comment in the form of a question.

"Yes, he does," Marcy said. "But he ain't drinkin' now, my father says. An' he can't help drinkin' either. He don't do it because he wants to. He tries an' tries to stop, an' he just can't. My father says we should all be sorry for him."

"He made all them little chairs for us," Stephen said. "An' he painted them, too, in all them different colors. I wonder what'll ever become of them little chairs."

Elly's eyes filled with quick tears. She had not sat in the little chairs so often as the other children; but two or three times old Mrs. Holt had invited her, too. Once she had even written a note and sent it up the hill by Thaddeus to invite Elly to a children's party. Elly looked at Marcy helplessly and saw to her relief that there were tears in Marcy's eyes also.

"My mother sat in them chairs, too, when she was a little girl," Marcy said. "Mis' Holt was awful old. She's been kind to lots and lots of children, my mother says."

"Do you always go barefooted on all these rocks?" Benny asked Elly, looking at her brown, dirty feet below her overalls.

Elly blushed beneath her tan.

"Yes, I do," she said with sudden courage. "What's it to you? I like to feel my feet on the rocks."

Marcy glowered at Benny, who looked frightened.

"We'd better go on to the swamp, Elly," she said. "We have to pick lots o' flowers."

3

They gave Davy another lift as they left the ledge and followed a narrow path through the woods toward the swamp. This time Stephen and Marilyn made the chair for him at Marilyn's suggestion. She felt uncomfortably that she was being left out of things, and she proposed the chair in an attempt to re-instate herself. The path was wet and dark beneath the overhanging trees.

"I've never been here before, I don't think, Elly," Stephen said, as he and Marilyn stumbled a bit with Davy. "You must know a whole lot o' places that we don't know."

"I reckon perhaps I do," Elly said, with an intoxicating sense of new importance.

After ten minutes they left the path and came out upon the edge of a small swamp set about by alders and great thickets of fireweed, now bearing the last of its fragile seeds. Even in the still air the wisps of fluff were slipping from their slender pods and drifting about in the sunlight.

"That's fairies' hair," Elly said.

"It's a pretty name," Marilyn said graciously. "How did you hear that name for it?"

"My father made it up," Elly said. "He thinks up names for a lot o' things that grow."

"Elly was right," Marcy cried, pointing excitedly to some late red wood lilies growing on the further side of the swamp and holding their bells upright above their sharply spiked leaves. "Seems like they just waited to blossom for old Mis' Holt."

All the children respectfully waited now for Elly's orders.

"Davy would best stay right here on this big rock," she said. "He might get in too deep. This swamp is a very wet place. I've saved you half my licorice, Davy. An', Benny, you stay along with him. See them blueberries there to one side? You can eat them. We won't be long, Davy, or ever out o' sight. An' when we get the lilies, we'll all rest awhile with you."

"All right," Davy said, reaching for the licorice.

"The way to pick the lilies," Elly said, "is to pull them straight up with their roots an' all. I'll show you how to do it. They'll keep better that way. Then we'll put them in a pool o' bog water, while we get some other flowers. Just across the swamp there's a sandy place where there's a lot o' sea lavender. We'll get that, too. I saw some once in a vase in old Mis' Holt's sittin'-room."

"Don't be afraid, Davy," Marcy said. "We ain't goin' far."

"I'm not afraid," he said sturdily.

"I'll keep an eye on him," said Benny.

When the four older children under Elly's direction had gathered all the lilies, which she carefully submerged in a pool of water among the green swamp hummocks, they broke off great masses of sea lavender, now less bright than in the summer, but still lovely with its delicate, mist-like blooms.

"It would be silly to get goldenrod here," Elly said. "We can get all we need of that on our way home."

4

They were hot, drenched, and tired when they had finished and returned to the big rock where the little boys awaited them, their mouths black or blue with licorice and berries.

"There's some speckled birds under them alders there," Benny said.

"They're thrushes," Elly said. "They're gettin' ready to go south for the winter. They all come to one place an' go along together. It'll be a long trip for them, way down south."

"How do you know so much?" Marilyn asked, still resentful of Elly's leadership.

"My father knows all about birds," Elly said. "He's got a book about them, but he don't need ever to look at his book. He just knows."

Marcy wiped her little brother's mouth again. Her handkerchief was black with the stains of licorice. She climbed down off the rock and sozzled it about in the swamp water. Then she spread it out on an alder bush to dry in the sun. It was a big handkerchief with red teddy-bears, wearing silly hats, upon it.

"That's a very pretty handkerchief, Marcy," Elly said.

"I'll give it to you, Elly," Marcy said. "My mother will wash an' iron it, an' then I'll give it to you to remember the day we got flowers for old Mis' Holt."

"You mustn't do that," Elly said, though her heart quick-ened with surprise and happiness. "I wouldn't feel right to take it."

Marilyn looked scornfully at the handkerchief drying on the alder bush.

"I've got six handkerchiefs at home," she said. "They're all different colors, and they all have *Marilyn*, for me, em-broidered on them. I got them for a prize in school."

This stupendous fact had a crushing effect on the other children, especially on Elly, who had no handkerchiefs of her own at all. Marcy leapt to Elly's rescue.

"Don't brag so much!" she said to Marilyn.

Benny also added his reproof.

"You're an old puff-ball!" he said to his sister. "You're always blowin' out your own fat face."

Stephen was conscious of the momentary tension and relieved, or deepened it, by a solemn question.

"Has anyone here ever been to a funeral?" he inquired politely.

The ensuing silence was proof that no one had.

"What do you do at a funeral?" Benny asked.

"You sit still," Marcy said. "We shall all sit still in the kitchen on them little chairs an' stools. We mustn't talk at all or move about. An' afterwards Mis' Norton is goin' to take us all with her into the sittin'-room to see old Mis' Holt there, so we'll all of us remember always how good she was to us."

"I don't want to see her," Davy said, beginning to cry a little.

"Then you shan't, dear," Marcy said. "Don't you worry. I'll take care of you, an' mother'll be there, too."

She put her arms around the little boy, who snuggled up to her and began to smile again.

Stephen felt once more that he would best change the subject, especially since he had been responsible for it.

"When old Mis' Holt went to sea in one of them big sailin' ships a long time ago," he said gravely, "she often saw whales, bigger than our houses, an' spoutin' water high in the air. She saw many icebergs too, great green icebergs like mountains floatin' in the ocean. An' lots of times she saw flyin' fish, an' great sharks, an' all other creatures of the deep."

"Did she ever see a sea-serpent?" Benny asked.

"Don't be silly!" his sister said. "There aren't any sea-serpents."

Benny felt a fierce alienation from his sister. At the moment he preferred any one of his companions to her.

"How do you know?" he asked her, angrily. "There are, too, sea-serpents, aren't there, Stephen?"

"There are *not!*" Marilyn cried. "I read it in my geography book in school. I learned the very words. It said, 'Contrary to popular opinion, there are no sea-serpents.'"

This unexpected erudition stunned them all, and Marilyn was quick to seize upon her advantage.

"And if you don't believe a book," she said loftily, "you can believe my teacher in Portland, Maine. She just laughs at sea-serpents."

No one could speak. The mention of such a faraway city as Portland was overwhelming. To all of them Portland spelled distance and mystery. They were awed into silence.

Elly was the first to recover. She suddenly recalled a piece of information far more exciting than sea-serpents.

"Thaddeus Holt was born at sea," she said slowly, waiting for the effect upon them all. "He was born off Cape Horn in a terrible gale o' wind. Old Mis' Holt told my father so, once when he was over there helpin' her."

The little boys stared at Elly, and even Marcy looked puzzled.

"That ain't true," Davy said. "There ain't any babies to be found for folks way out to sea, are there, Marcy?"

Marilyn flashed Stephen a knowing look, clapped her hand to her mouth, and snickered. Marcy turned upon her furiously.

"Don't you dare to laugh at him!" she cried. "He's just a little boy. You're the meanest girl I know in the whole world! Don't you mind her, Davy. We'll ask mother when we get home."

"I don't have to ask my mother," Marilyn said haughtily.

Stephen, moved by the sight of Davy's round, perplexed eyes and Marcy's anger, felt suddenly a strange weight of sorrow for them both. He silenced Marilyn with a cold, calm stare of disdain. Then he turned to Elly.

"We was all little once, like him," he whispered to her.

5

Elly, from whom, as in the case of Stephen, life had concealed little, given their own particular surroundings, was sorry that she had made her startling announcement since it had seemed to separate her from Marcy. She tried now to weld the group together by the exercise of her original authority.

"We'd better be goin'," she said. "It's a long way to old Mis' Holt's. Marcy an' I will help Davy, an' Stephen will carry the lilies, because he'll be careful of them. Benny an' Marilyn can take the lavender. When we get out to the road, we'll get the asters an' goldenrod, an' then we can pick some more of them common things right in Mis' Holt's field."

They followed the path silently back, past Daniel Thurston's, where Stephen and Elly saw to their relief that both his boat and his dory were floating on the incoming tide, and on through his wood-road to the hill above the cove. There they rather instinctively separated into pairs. Marcy held Davy by the hand. Marilyn and Benny began to break off tall stalks of goldenrod and white and purple asters.

Stephen and Elly, each feeling curiously drawn toward the other, divided the lilies and led the way down the hill.

"That scow's awful big," Stephen said anxiously. "It will want a heap o' flowers."

"Don't worry," Elly said. "Mis' Norton says they'll have the scow hauled right in alongside Thaddeus' slip. You an' I could stay after the other children have gone home to dinner, an' begin to put our flowers on it. There are heaps more growin' close by."

"That's a nice plan," Stephen said. He hesitated. "You goin' home to dinner, Elly? Your folks goin' to the funeral?"

"My folks had to go off early this mornin'," Elly said. She tried hard to make her words sound at once regretful and important. "What about your mother? Is she goin'?"

"I wouldn't know," Stephen said. "She might not even be home, like your folks."

Elly was quick to sense his anxiety. All at once they two seemed alone, far away from the others.

"When we get the scow ready," she said, lowering her voice, "you an' I'll run along the beach an' up the path across the headland to my house. I can get some dinner for us both."

"Shag Island looks a lonesome place," Stephen said, after a few moments.

"I don't mind lonesome places," Elly said. "I like them."

6

Marcy now decided to rejoin the group. She had picked her arms full of asters and had given some to Davy to carry. She had decided, too, that she liked Elly Randall and wished to have her as a playmate, once she had convinced her mother of how nice Elly was and how kind she had been to Davy. Marilyn and Benny now came from the roadside thickets. Together they all walked down the hill.

"Shag Island don't look so lonesome when you go there for a picnic," Marcy said. "We went once. I wouldn't promise anything, of course, but my father just *might* take us all out there some day soon in our new boat if I asked him hard enough."

Elly's heart seemed to leap into her mouth. She had never been on a real picnic since she had come to the cove. Marcy's almost careless words, which embraced them all, sounded unbelievable to her. Suddenly the harbor with its boats, now in and at anchor, the snug houses lying below the hill, Shag Island with its dark trees, the funeral where they would all sit together on the little chairs, seemed to offer her new and priceless gifts. She thought she would like to stay here for always.

Now she hardly dared speak for fear that Marcy had been cruelly joking.

"You really mean it, Marcy?" she said, in fear and wonder.

"Sure I do," Marcy said, expansive and proud.

"You mean *us*, too?" Marilyn asked. Benny was staring at Marcy in doubt.

"Sure," Marcy said again, her triumph rendering her magnanimous. "If you don't put on airs again about Portland, an' if you ain't mean again to Davy or to Elly. I said all, didn't I? Well, I mean all."

"My father's boat is the best boat on this whole coast," Davy said. "He's goin' to let me steer her when I'm seven."

Stephen shifted the lilies so that he could take Davy's hand to keep him from stumbling on the steep, rocky hill.

"You can pick more asters if I look out for him," he said to Marcy. "We can't have too many for that big scow."

He looked at Elly.

"If we just *should* go on that picnic to Shag Island," he said, "we could gather some more flowers right out there to put on old Mis' Holt's grave."

(198)

PART
THREE

The Funeral

❖ ——————————————————————

The Funeral

◆ ─────────────────────────

W ITH ALL her gratitude and admiration for Sarah Holt over these many years, Lucy Norton was always sensitively aware that there were thoughts in the older woman's mind which she could never wholly grasp, areas which she would never enter. Surprised by some chance remark of Sarah's or perhaps just by seeing from her face that she was dwelling alone in some enclosed, yet uncharted space, Lucy wondered about this separation from others. She asked herself whether, if she had perhaps lived at another time and known all the peoples and places which Sarah had known, she, too, might have been someone quite different, able to think far deeper thoughts, to understand people better, and to escape anxiety and fear. She could, of course, never answer this question or the more profound ones arising from it; and Joel, to whom she had put them occasionally in their earlier years at the store, was not in the least enlightening.

"If you can't fathom it, Lucy," he said, "you surely can't

expect me to. You are much smarter than me. Folks are just folks, I guess, each one different from the other."

This answer did not satisfy Lucy. For the differences in Sarah Holt were not mere differences or dissimilarities in abilities and traits, like, for instance, the difference between herself and Joel, she being quick and Joel slow, or the difference between Sam Parker and Ben Stevens, Sam being kind and genial and Ben crabbed and silent. These were but superficial and obvious differences which one saw merely by looking about among one's neighbors.

Sarah Holt was different in quite another way, Lucy knew, trying hard to discover just what this way was, until finally she came to realize, acutely if not completely, that Sarah was different from anyone whom she had ever known because of the singular manner in which she looked at things, because of the decisions she could make, because of the questions she could ponder over as though they were not intimately connected with her at all. Unlike her neighbors, she never seemed to become involved in trivial affairs, messed up in them, worried over them, even controlled by them. She was always in some odd way outside them, observing them, holding them far off so that they had no power to nag at her as they nagged at others. In comparison with all other people in Lucy's world, she seemed unfettered and free in her mind. It was this quality which made it impossible for one to feel sorry for her, growing old there with Thaddeus, watching with pity and understanding his desperation and failure, sending away his wife and son so that they might not be forever broken by his hopelessness.

Often in the years she had known Sarah, Lucy had been

startled and even shocked by this way of looking at things. There was, for example, the night that Thaddeus had almost died from drowning.

This accident had happened shortly after Nan had left for good and some months during which he had not been drinking at all. He had become another person in those months, filled with ambition, with plans for the future, proud of his place in the community. He had joined the other men in the store in the evenings, conscious of regaining their respect and confidence. You could never have recognized Thaddeus Holt in this new dignity, all the men said.

As if to further his fresh hopes and give them realization, a great rush of herring filled his weir. While he was seining it at night with the help of his neighbors, for the tide was right only then, he missed his balance in his dory and fell into the water, getting entangled in a net as he did so. When they finally got him free, he was so nearly dead that it took fully an hour to bring him round again, the closest shave, the men said, that they had ever seen, and they had seen plenty.

"I wish they hadn't brought him round," his mother said to Lucy a few days afterward. "If they hadn't, he could have died while he was still proud of himself. Now he probably can't manage that. It would have been wonderful for him to die with everyone respecting him and with him hopeful. I couldn't have asked for a better thing for him or for me. Just the right time for him to die won't likely come again, more's the pity."

Lucy had been stunned into silence by these calm words and had guarded them carefully against the greater shock which they would have given to others. She was more stunned when she found herself believing that Sarah Holt had been right in saying them. They kept ringing in her ears during the weeks and months that followed, when Thaddeus, as if in fulfillment of his mother's somber prophecy, began to drink again.

She had been shocked, too, though in a different way, when, on Sarah's ninetieth birthday only a month ago, she had told Lucy her desires and plans concerning her own death. Lucy had made a cake to celebrate the birthday; and they were having a party quite by themselves in Sarah's sitting-room.

"Ninety," the old woman said. "I really never thought I'd make ninety, Lucy. I wouldn't mind a few more years, with you here and so many things to think about and to see. I've shed a lot of tears over life as it is, but not many over my own life. I haven't the least wish to quarrel with that. We may as well face facts, though. I'm not exactly young any longer and since, with things as they are, I can't always depend on Thaddeus, I guess I'll have to bother you with a few matters on my mind."

Lucy took off her glasses and wiped them carefully. She was embarrassed, proud, and frightened all at once.

"Where are you planning to be buried, Lucy?"

Lucy hesitated, as she replaced the glasses.

"Why, I don't just know," she said. "Somewheres side of Joel, I s'pose. Perhaps someday we'll go back home to the island to live. We sometimes talk about it."

"Well, the fact is," Sarah Holt said, "I don't in the least want to lie beside my husband in the old Holt burying-ground up there in the field. Wherever he is, he isn't there, and, except for Tom, I never much cared for the Holts. That's not the reason, though. I've always had the notion that if people could manage it, they ought to go back to their beginnings, the places where they started from. I suppose it's just the idea of rounding out the circle of one's life. The world's got so now that there isn't much chance of ending your life where you began it. People can't put down roots any more with much confidence that they're going to hold firm. Perhaps it will turn out to be a good thing in the end, but I'm not so sure about that.

"I've been thinking a lot lately of just why this notion of a circle appeals to me. I suppose perhaps it comes from the sea. You always felt, when you sailed, that you were completing a circle. If you went to the China coast, you could approach it from the east or from the west courses, and whichever way you sailed back home, you always completed a circle. It's much the same way with the tides rolling in and out. Or with the horizons one used to see on every side in open water. I suppose you'll think I'm crazy, Lucy, when I say I'd like to be taken back to Shag Island and lie there where I began to live."

"I'd never think you was crazy," Lucy said slowly, at a loss for words. "But Shag Island's become an awful lonesome place."

The old woman smiled.

"I won't know a thing about that," she said. "It's where I came from and where I'd like to be taken, that is if it could be managed without a lot of fuss. You might tell Joel and Sam Parker some day, and I'll tell Thaddeus if the right time comes." She smiled again. "I can't help thinking," she added, "of all the flutter I'll be giving to everyone

(206)

around. Things won't be dull for this village on the day of my funeral, Lucy."

2

Lucy thought of these words as, once she had pulled herself together and been able to collect her scattered wits, she hurried about her many chores. There would be no trade at the store, she knew, for everyone, now that the boats were in, would be making ready for the funeral, getting a hurried dinner out of the way, putting on their best clothes, and starting forth in a flurry of unwonted excitement. Nor could she honestly say, with all her sorrow and loneliness, that she herself was not excited, what with the beautiful day, the children's finding the lilies, the knowledge that these extraordinary arrangements were in readiness and planned to the last detail, and the sense of a common and consuming interest uniting all.

She had just finished tidying the rooms upstairs, inspecting Joel's blue suit and ironing his white shirt, when he drove into the yard with his load of provisions. She ran downstairs to help him unload. He looked tired and anxious. She was careful not to ask about the frankfurters, but was vastly relieved when she felt their spongy lengths in a generous brown-paper parcel. They worked quickly in their familiar, long-established routine, he unpacking the bales and bags, crates and cartons, and she arranging their contents in order upon the shelves and within the counters. The tired lines left his face as they worked, and when they sat down to dinner, he was himself again.

"I gather quite a lot of people are comin' from away," he said, "with the day such a good one and she so respected by so many. An' all the backwater folks aim to come. Most everyone stopped me as I came along to make sure o' the time."

"That's what I planned for," Lucy said. "The parlor's all ready and the bedroom, too, and if there's too many for the house to hold, there's the field outside. It's lovely there in all this new sun."

Joel looked gratefully at her.

"Is Thaddeus all right?"

"Fine. He's just fine, only sad and quietlike."

"I'm sorry for him," Joel said. "No one mustn't be unkind, no matter what he does now that he's all alone."

Lucy thought for a few minutes. Then, having made up her mind to speak, she said:

"Did you ever think, Joel, that it might have been better for Thaddeus if you and Sam hadn't worked so hard to bring him to, that night when he fell into the weir?"

Joel stared at her. He looked scared, as he always did when he was taken unawares by anything at all.

"Men can't do things like that to each other, Lucy," he managed to say at last. "They've got to do the best they know how to do when the time comes."

Lucy wished she had not spoken. Joel seemed suddenly as wise to her in his own way as Sarah Holt had seemed in hers.

"No, of course, they can't," she said quickly. "Yes, of course, they must, Joel."

For she knew now that Joel was right just as Sarah had been right. It was one of the strange things about life, she thought, that different people could think in such different ways, and yet both be right.

3

When they had stopped by for Sam Parker and were all three in Joel's truck going toward the Holt place, Lucy felt proud of her men folks. There was hardly any shine on Joel's blue suit. He was freshly shaved, his shirt and tie completely right, his black shoes polished. Sam, too, had taken great pains with himself, though Lucy thought he looked sunburned and tired. He had set great store by Sarah Holt, she knew, and he would miss sadly all his hours with her. Lucy herself was in her black dress, which had done valiant service now for a long time; but she had managed to refurbish her old hat with a new black velvet bow.

"The tide's just as it should be," Sam said, as they neared the head of the cove. "I don't know, Joel, whatever we should have done with the tide wrong. We just couldn't have managed things as she wanted them if the tide hadn't favored us this way."

"Seems as though everything has worked out just right," Joel said.

Lucy, sitting between them on the high seat of the truck, felt all at once alone, just as hardly a moment ago she had felt proud to be with them, strengthened by their nearness. Neither of them could understand just how right everything was, she thought. She was swept again by the realization of those strange moments when she had stood in the sitting-room looking down at Sarah Holt, when, all confusions stilled, the meaning not alone of Sarah's life, but of life itself as it was lived on this small piece of barren earth by this remorseless sea had been revealed as ennobling and

of great price. When the late autumn and winter crept on
and this frozen road thudded beneath one's feet like plates
of iron, and gales swept in with the first sleet and snow, she
would look back upon this day as to some lighthouse tower,
rendering harmless to those within its glow the sea and the
fog.

Now she saw her neighbors from both cove and backwater coming from their homes, all dressed in their best,
all silent and expectant. For this funeral, as its provider
had prophesied, was bound to be different from any funeral
they had ever attended; and, try as they might to conceal
it beneath their good clothes and solemn demeanor, curiosity was competing successfully with regret and grief. And
because Lucy Norton was three-quarters one of them and
only one-quarter a thinker or a mystic (and those endow-
ments not so much indigenous, as transplanted) she, too,
coming suddenly back to earth, found herself no longer in
a realm of abstraction, but sitting most concretely between
Sam and Joel, hoping that her black dress was not getting
too creased in the narrow space of the front seat, and, like
everyone else, quickened by the pervading atmosphere of
suspense and drama.

Joel stopped the truck at the head of the cove to pick
up Stephen West and Elly Randall, who had just then
scrambled from the high beach into the road. Both were
freshly scrubbed, in clean clothes, and serious.

"I see you got the lilies," Lucy said, turning to smile at
them as they stood, holding onto the back of the seat.

"Elly knows everything about where flowers grow,"
Stephen said.

Elly turned away her eyes to look at the cove. She felt

ill at ease to be driving to old Mrs. Holt's funeral. She would much rather have walked on with Stephen. She nudged him now and whispered:

"There's Dan comin' across the cove in his dory."

Joel did not start up the truck. They were all staring at Daniel Thurston, standing up in his dory, pushing it forward across the still, full tide with long backward strokes of his oars. He was dressed up, too, even to his black hat. Between the steady sweeps of his oars he stood, tall and firm, stooped though he was. He and his boat were the only moving things in the quiet cove where the fishing-boats lay at their moorings, and the skiffs and dories floated placidly near the shore.

Joel was distressed to see him out there pushing himself across the half-mile of water.

"I'd have gone up for him," he said, "if I'd known. I thought he was sick an' couldn't come."

"Daniel never stays sick long," Elly said, finding her voice.

"He don't look right," said Stephen, "without his dog, lookin' ahead there in the bow."

Lucy was distressed, too, by his solitary dark figure coming on and on. She was also perturbed about his chair in the sitting-room.

"Is your mother coming to the funeral, Stephen?" she asked.

"No, she ain't," Stephen said. "Leastwise, she's not to home."

Lucy counted silently on her fingers. "Thirteen," she said to herself, relieved. "That's exactly right."

4

The field around the old Holt place was generously dotted with people. There had not been so many people at the cove since she could remember, Hannah Stevens whispered to Ben. There must be at least sixty, she said. Those who came "from away" or from the backwater left their cars or trucks in the pasture, directed to places there by Carlton Sawyer, whom Lucy had assigned to this responsibility. After they had left their cars, they joined the cove people in the field before the house.

All stood about in embarrassed silence, waiting for the signal to go indoors. The fishermen would have liked to smoke their pipes, but they refrained from doing so. They felt strange out of their oilskins, corduroys, and hip-boots. They stared at the beach where a scow was drawn up by Thaddeus' slip and at his fishing-boat moored just beyond. They studied the tide, due to turn before three o'clock and even on this still day likely to prove a nuisance, if not worse, to a towing job there at the mouth of the Tidal River. The women stood beside their husbands, though they would have preferred to be with one another and to make a whispered comment now and then. They looked curiously about, absorbed in their own thoughts. They gazed at Shag Island with its dark spruces and at the autumn flowers which banked the inside edges of the scow with a touch of crimson showing here and there among them. And everyone in the little company, men and women alike, stared at Thaddeus Holt, standing beside the scow and talking with Joel Norton and Sam Parker.

When Daniel Thurston reached the shore in his dory

and Sam Parker had given him a hand at landing and at beaching his boat, he, too, joined the silent people in the field. He walked slowly up the grassy slope, carrying his black hat. Carlton Sawyer brought a chair for him, but he disdained it, leaning instead against a tree until he had wiped his streaming face with his handkerchief and put on his hat again. He had not had on his old black suit in years, or his heavy gold watch-chain, which looped across the buttons of his vest, or his starched shirt and collar. Nor had he, in fact, been for years to the Holt place, since one day, at least a decade earlier, when Sarah Holt had taken him severely to task for certain of his ventures in the neighborhood and along the coast. But as the oldest resident at the cove now that she was gone, he felt a strong sense of convention, even of propriety, and he was here from duty as well as deference. Everyone looked at him and not without a certain respect, reluctant though it might be. He was well past eighty, they all knew, and his own time could not be far off.

The children sat together in a little knot on the grass near the front door, all scrubbed and brushed, in clean frocks and blouses, all shy and solemn. Marcy held her little brother by the hand, every once in a while assuring her mother and father with a glance and nod that she was looking out for him. Elly happily shared her responsibility.

"When we go into the kitchen," she whispered to Marcy, "we'll put him in the little red chair between you and I. That chair's got arms on it in case he gets tired and sleepy."

"Okay," Marcy whispered back.

There was a visible stir among the people in the field when a small car came down the pasture road and drove off into the grass as though its owner were well acquainted with the place, and at home. A woman got out of the car and walked toward them. She was tall with gray hair and a thin, quiet face. She nodded with a half smile to those whom she passed. Most were too startled or too curious to return the smile even if they had been minded to do so. She went toward the front door where Lucy Norton was standing, greeted her and apparently asked a question. Then in the midst of stares and side glances and general amazement she walked toward the three men at the shore, who were just turning about to come up the slope to the house.

Sam Parker and Thaddeus took off their hats when she reached them, as did Joel after a few awkward moments. What she said to Thaddeus every woman watching from the field and most of the men would have given their all to know, for at this agitated minute life had driven death completely from their minds. But they could hear nothing, and, as Hannah Stevens regretfully thought, Sam Parker and Joel Norton were the tightest-mouthed people in the place. When those at the shore started toward the house, Nan Holt slipped her arm within that of Thaddeus, and they walked up the slope together.

No one was disappointed or disturbed that the funeral service did not begin until nearly three o'clock, except the men who were in charge of the scow and who worried about the tide. All the others were far too curious and expectant to regret the delay. As the time went on, the silence and solemnity gave way to low conversations and to moving about from one small group to another. Nan and Thaddeus Holt stood apart from everybody else, talking together and seemingly unconscious of the commotion they were causing in most minds. Mary Sawyer brought Daniel Thurston a glass of water, and Carlton at last induced him to accept a chair. Nora and Seth Blodgett came to stand beside him. Lucy brought the children some cookies which they munched silently. And as everybody waited in the field, the tide filled the cove; the sun moved westward, touching the wings of the gulls as they settled upon the water or circled slowly above the headland; the air hummed with the high whine of the crickets and locusts; Shag Island was flooded with afternoon light; and not a cloud was visible in the brilliant September sky.

"I don't know as I ever see so pretty a day on this coast," Daniel Thurston said to the Blodgetts.

"It was even prettier when Seth and I went up river this morning," Nora Blodgett said.

Lucy shared the concern of the men who had the scow in charge. A line from somewhere kept running through her head to the effect that the tide waits for no one; and yet she was unwilling to begin the funeral service without the doctor. She knew he would come, no matter how many patients he had to leave in his office. At a quarter before three his car hurtled down through the pasture road and into the field. He had brought some of his roses, and just before Lucy went to the front door to summon everyone within the house, he and she placed them at the foot of the coffin.

5

Both Lucy and Joel Norton had been more than a little anxious about a minister to conduct old Mrs. Holt's funeral service. Such a problem was never a slight one among detached, unfrequented communities where distance and long winters in themselves prevented people from any regular attendance upon churches, and, therefore, from adherence to any particular parish. The Seacoast Mission, to be sure, did its utmost along remote portions of the coastline and among the islands to afford help of every sort, in sickness and in health; but with hundreds of miles to cover with its one boat, its services were not always immediately available. The normal religious activities of even a small coast town or village were unknown among those who lived at the cove, except to Hannah and Benjamin Stevens; and their peculiar brand of spiritual consolation and expression did not command the marked respect of their neighbors. To such far-flung fishing settlements ministers could rarely be essential in life, and they were by no means indispensable in death. During her thirty years at the cove Lucy had many times gone with Joel in the truck to some neck or headland, or with Sam Parker in his boat to some island, where there was no one to say anything at all to the few people gathered in some old family burying-ground or in some small, disheveled common graveyard, once perhaps in the center of a lively community, but now lonely and deserted.

She could not bear that Sarah Holt, who for much of her life had known a far different world from this one at the cove, should at her death be at the mercy of its isolation, even although she felt certain, and perhaps uncomfortably so, that a minister to conduct her funeral would have meant little or nothing to Sarah, whose expressed wishes had concerned only her place of burial. No one among the neighbors, except for the Stevens (and Hannah now had her secret misgivings), was overly eager that young Mr. Simpson should officiate. Thaddeus during his mother's brief illness had been in no state to give advice, and after her death had seemed quite indifferent as to the manner of her funeral. But since ministers were both distant and scarce and Mr. Simpson willing and even eager to come, once Joel on his way to town had asked his help, he now stood a little uneasily in the sitting-room with his Bible in his hand.

To Lucy's surprise, Thaddeus, once he had learned about Mr. Simpson, had himself decided upon what the minister should read for his mother. She had found him the afternoon before the funeral sitting at the kitchen table and quietly selecting the passages which he thought best.

"Just tell him to read these," he said, after he had finished and carefully copied his selections on a sheet of paper. "My mother wouldn't wish him to say anything, I'm sure, as ministers sometimes do at funerals."

Mr. Simpson was not doing badly at all, Lucy thought, sitting in the chair nearest the kitchen door where she could look around at the children. They were all quiet and good with their hands folded, she saw, though they looked subdued and uncomfortable. She was glad she had placed their chairs and stools in a circle for them. Every few minutes she smiled at them, and Elly and Marcy returned the smile shyly as if to say that they were seeing to things in this once friendly room, now become so strange and solemn.

Lucy found her mind wandering during the minister's reading, much as it had wandered earlier that morning. All the things which really mattered had happened in those few brief minutes when she had been alone with Sarah; and the words which he was now reading were like some unimportant postscript to a most important letter.

> *They that go down to the sea in ships, that*
> *do business in great waters,*
> *These see the works of the Lord, and his*
> *wonders in the deep.*

Sunda Straits, she thought, with their reefs and shoals and cross currents; the Kerguelens, rising from a waste of waters; the gray, perilous, tossing seas off Cape Horn. That boy with his tears, which the bos'n wiped away, and his fiddle, and the ropes aloft catching the tardy wind. When the clock struck three, she heard it again, striking suddenly seventy years ago in the dim hold of Captain Holt's ship above the tumult of the storm.

She turned to smile at Davy Sawyer in the little red chair. She was afraid that the clock had startled him.

> *The days of our years are threescore years and ten, and if*
> *by reason of strength they be fourscore years, yet is their*
> *strength labor and sorrow; for it is soon cut off, and we fly away.*

Not always labor and sorrow. Sometimes gayety and defiance and strength for others. She looked along the two lines of chairs. Daniel Thurston's folded hands across his black vest were shaking too much. Someone must drive him up the hill and then row his dory across the cove for him; and she must see that he had something for his supper.

> *Man goeth forth to his work in the morning, and to his labor*
> *until the evening; for the night cometh wherein no man can*
> *work.*

Or women either, she thought, only they did, and often far into the night. The Bible always seemed to favor men above women. She looked along the chairs to the women sitting there, and in her mind they became multiplied to all women in places like this, in the backwaters along unfrequented roads, on other shores and islands—rising before dawn, getting their men off to their boats, spending their days at common, threadbare tasks, anxious over winds and fogs. A great compassion for them all filled her wayward mind.

If I take the wings of the morning and dwell
in the uttermost parts of the sea,
Even there shall thy hand lead me,
and thy right hand shall hold me.

The wings of the morning. Those were pleasant words. Did they mean the sea-birds that flew at dawn, gulls and terns, kittiwakes and hawks, shags and herons, their wings touched with early light? Or did they perhaps mean clouds at sunrise?

From where she sat she could see Thaddeus' bent shoulders in the exact center of the front row of chairs. Tears filled her eyes as she saw that Nan Holt, who sat beside him, had placed her arm again within his.

And I saw a new heaven and a new earth, for the first heaven
and the first earth were passed away. And there was no more sea.

However would Sarah Holt manage if there were no more sea where she was going, or had already gone? The one who had written those words must have fared ill from the sea since he had wanted none of it in his new heaven and his new earth. He could hardly have known such a day as this one when the sea gave back all its gifts a thousandfold to those who lived beside it.

Davy was getting restless, she saw, and beginning to cry a little. It was about time that all the children went out-of-doors. They could even pick some more flowers right here in the field along the head of the beach. It would be better after all if she did not bring them in to look at old Mrs. Holt.

Just as the minister began his prayer, she slipped into the kitchen, closing the door to the sitting-room.

6

After the cove fishermen had carried the coffin down through the field, placed it upon the scow, and heaped the children's fresh asters and goldenrod upon it until it was quite hidden beneath them, the mourners at this strange funeral gathered along the beach-head and the narrow strip of shore. Tact and precedence determined their positions. The few from the small towns and villages along the highway, who had come out of respect for Sarah Holt's long life and who for only this brief hour were a part of the cove, stood in the background upon the high edge of the field. The backwater men and women in their turn took their stand at the head of the beach, giving place to the immediate neighbors, who clustered together just above the tide.

As she gathered the children about her in an awed little group and anxiously watched the men who were now maneuvering Thaddeus' boat into position and attaching their long ropes to the scow, Lucy Norton was seized by the fancy that somewhere she had heard about ancient

barges in far-off times and countries that once had borne
the dead out to sea after a like manner; but she was too
perturbed just then lest things should go wrong to
straighten out such vagrant thoughts. Tides did not serve
for funerals or for anything else except their own relentless natures, however gentle they might seem to be; and she
felt a sense of dread hovering about her which she knew
was shared by all those long familiar with their ruthless
ways.

She thought she could never be grateful enough to Hannah and Benjamin Stevens or even think unkindly of them
again, when, just as Thaddeus started his engine and the
scow grated against the slip, they began to sing some old
hymns that everyone knew, "Shall we gather at the river?"
and "On the other side of Jordan." Everyone joined in,
even the visitors and the backwater people, until the first
swirling of the scow in the swiftly running channel had
been overcome by the tightening of the ropes in the skillful hands of the men, and the even, sure throbs of the engine, now clearly equal to its task, proved that this tide
at least had been beaten and that all was well.

After the singing stopped, the people in the field and on
the beach-head still stood quietly on, unwilling to leave
until distance and the late afternoon light should blot out
this impressive spectacle, moving ever farther out to sea.
When there was finally nothing visible except a blur against
the sunlit trees and high ledges of Shag Island, they moved
up the slope toward their cars beyond the house, leaving
only the small group of neighbors on the shore.

Perhaps these continued to linger still a few more minutes
by themselves because they were conscious that this brief

interlude in time would not come again, this sense of unity, almost of entity, which was binding them together. Now they were forgetful of the anxieties and necessities which wove the simple pattern of their lives. For these brief moments they were dimly aware of deeper, nameless feelings, of restless, mysterious forces quickening their hearts, clearing away the confusion within their separate minds. Tomorrow would be different. They would be drawn apart tomorrow, each intent upon his own existence, each caught and held within his own snares, woven from the past or constantly entangled by the present: Nora Blodgett with her frail, new hope; Daniel Thurston with his old age and loneliness; Hannah Stevens' unwritten letter; Mary Sawyer's fearful wonder as to how long love might last. The questions asked tomorrow would be the paltry questions of their own small place and time, not these strange, half-recognized questions about the round of life and one's meager part therein. They would again become involved in the certain loss of Thaddeus Holt's fleeting stature, in his wife's inexplicable behavior, in the suspicions aroused by the temporary, shifting members of their community, who today had made clear their alienation from it by their absence. In the ordinary, persistent routine of days, what they had seen this afternoon would become little more than a stirring pageant, and the old woman who had granted them a dignity and value they would never have accorded to themselves, a fast-receding memory.

Lucy Norton, standing among them, saw them all and herself with them as Sarah Holt had taught her to see them. Now if the doctor asked her why she had lived here for thirty years and why at this moment she knew she would remain for the rest of her life, she could perhaps tell him, although he would never ask because, like her, he, too, had been given his moments of understanding. She wished now like everyone else, though perhaps more clearly, that they all might stay like this, untroubled, unafraid, careless of the winter, in touch with some Reality which redeemed their lives and gave them worth. But the children were becoming impatient; the store was filled with her backwater neighbors; the sun was already disappearing behind the headland; and the edge of darkness was creeping on.

Nan Holt was the first of them to leave. She went quietly across the field to her car, and in a few minutes she had turned from the pasture into the road which led to the distant highway.

7

Trade had never been so brisk at the store since Lucy could remember. Not only were all the backwater people there, but all her immediate neighbors as well. Everyone had come straight from the funeral, partly to replenish their depleted pantries and neglected larders, partly, she felt sure, to preserve the sense of dependence one upon another. They crowded the small floor space and stood about on the porch, cove and backwater people together. After their shopping was completed, they formed into little groups in the yard, talking quietly in the fading light.

Lucy had no time to take off her black dress or even her hat, or to go upstairs for an apron. She was apportioning the frankfurters, which everybody wanted, weighing potatoes and onions, cutting out triangles of cheese, slicing bacon into innumerable strips, and hoping almost against hope that the bread would hold out. She dashed from one counter to the other after a spool of thread here, a bottle of liniment there. She added up long columns of figures which would have proved Joel's complete undoing, made change frantically, and now and again extended cheerful credit. She found time for a word with all the backwater people, whom she wanted to make feel at home, for they were rarely all together like this.

"I was awful relieved to see you right there, Henry. I knew in case the tide got nasty, you'd know just what to do."

"I really can't believe you made that dress, Alice. It might have come straight out of the catalogue."

She snatched a moment to carry out a bag of jelly-beans to the children who, like their elders, could not conceive of going home. She was more tired than she had ever been in all her life. She was also more happy, and contented in her mind.

Hannah Stevens was a bit later than the others in reaching the store. She brought half her morning's cake to Lucy for the men when they should get back in from Shag Island. She was saving the rest for Mr. Simpson, she said, whom she and Ben were keeping for supper. When she saw how Lucy was having to hustle and what a whirlwind things were in, she set to work herself with the best of will and a pleasant word for everybody. She even seemed oblivious to her grandchildren outside with the others.

"The singing was lovely, Hannah," Lucy said, nervously counting the remaining cans of condensed milk while she sliced some salt pork. "I never would have thought of that, and it came at just the proper minute. I didn't realize before how well Ben could sing."

"Well, it sort o' took the strain off folks," Hannah said, "what with all the worry about the tide. Mr. Simpson was real pleased, too. He was feelin' all along, he said, that there ought to be something religious down there to the shore. Let me slice that salt pork, Lucy. You're makin' an awful mess of it."

Lucy relinquished the knife and with relief gave her full attention to the condensed milk, the demand for which was really frightening.

"You ought to run a store, Hannah," she said. "You're a heap handier than I am at things."

The rush of trade was over in half an hour, though it had really seemed, Lucy thought, half a lifetime. And the reason for this curious extension of time lay in the fact that people without knowing it had suddenly opened the doors to their lives instead of closing them tightly as they usually did. Even while she had packed cartons, and added up money, and rushed all over the place in her best clothes, she had the odd and comfortable feeling that she was entering through those doors to sit for a few moments in the midst of those lives. With Henry Nevells, who, everybody agreed, had been the best navigator within miles of sea until the Coast Guard caught up with him and with certain of his stealthy talents. With Alice Staples, who, rumor said, was anything but a credit to the backwater with her two unverified children, but whose face had brightened like a child's over praise of her dress. Can people make places, she asked herself, finally snatching a moment to take off her hat, or do some places make people, in spite of all that they can do?

She went out-of-doors, for now with the early September evening falling across the cove and the stars about to come out, everybody was at last making ready to go home, either to their houses above the harbor, which after all were close together, or to their widely separated ones in the backwaters of the long road. Just before all got underway, driving or on foot, a car drew up at Trudy West's mailbox, where it left her, and then swept by up the long hill. Several of the people about the store raised their arms in the familiar coast greeting.

It was a long time, Jim Randall thought, since anyone had waved to him.

8

When everyone had gone, Lucy Norton locked the cash drawer, not without a satisfied glance at its generous contents, put the key behind the cans on the highest shelf, made a careful note on the crying necessity for more condensed milk, and went upstairs. There she took off her black dress, got into a clean cotton one, built up the fire, put the kettle on, set the table, and lighted all the lamps. When the men came in tonight, they would have not only the stars above them, but the lights of the village houses as well, all visible from the moment they got underway from Shag Island. Before she had left the Holt place, she had lighted a lamp there against Thaddeus' return, and now she would run over and do the same for Sam Parker. Seven lights did not seem very many, to be sure; but, spread out along the shore with the dark hillside behind them, they made a quite considerable illumination.

Once she had come back from Sam Parker's, she put on her coat against the chill of the September evening and sat down on the porch steps. Everything was still. The tide was now far out. She suddenly became aware of the flashing of the great light. She had been too busy to watch as she usually did for its first rays just at twilight, and in her concern for the lesser lights in the houses, she had for the moment forgotten it. There it was, straight out toward open sea, just as it had been, not only for her thirty years, but for nearly a hundred more. It threw its gleams every four seconds; to the south, over vast stretches of open water; to the west, over the bays and estuaries; to the east, over numberless points and islands; to the north, over their own cove and hillside. When the men left Shag Island, it would light their way as, four miles farther out, it sent its intermittent flashes behind their path homeward.

She was still tired. Her back ached from her neck to her thighs, and the calves of her legs were as taut and stiff as anchor chains. But she had been tired before, and yet quite without this sense of contentment, even of serenity, which now enfolded her. She thought she ought to be filled with sorrow over her loss; or with regret that days could never again be as they had been; or even with terror that now she had no one who could straighten out her muddles and quiet her fears. Yet, sitting here in the early darkness, she could not summon up one of these emotions. Everything now seemed right and clear. The wide circle of Sarah Holt's life was being closed, out on Shag Island, just as she had wished it closed; and this narrower, still open circle, which outlined her own life and that of her neighbors, had tonight become both right and even desirable.

She was surprised now at the nature of her thoughts. She had sat down here in the stillness with the intention of reviewing in her mind this extraordinary day, of trying to recapture the strangeness of the morning, all those fleeting perceptions she had known; but its events were eluding her. She could not lay hold upon them. Instead she was seized upon and engrossed by the present, not by the past or by the future. Contrary to all her plans and expectations, even to her desires, what was concerning her was the familiar, changeless, reiterative round and routine of the days as she knew them, days spent on this remote spur of land where she had lived for thirty years and where she would go on living until she, too, closed the circle of her life.

Here they all were, she and her neighbors, a handful of obscure, inconsequential people, working from dawn until dark in painstaking, unremitting labor, always vigilant and often anxious, drawing their means of existence from the sea, which was far more often their enemy than their ally. Almost no one knew about them. They and others like them, in scores of similar communities along the coast and on its islands, counted for little or nothing in the minds of those who lived in great cities, or on the vast plains, like Jeff, or in the mountains.

Even the hosts of summer people, who were now affording the coast its most lucrative industry, knew little about that coast during their three months of sojourn. They cruised about its bays and islands for rest, or health, or pleasure, thinking only how beautiful it was. They did not know it in its winter storms and cold, with no sun and warmth to conceal its jagged outlines, its jutting headlands, its cruel tides. They would return to it only when it had withheld its forbidding isolation for their brief season. Their gay boats filled its safe harbors ready for their races and their picnics. Few, if any of them had ever seen its black, tossing waters in November when other boats still put out to sea, not for pleasure, but from necessity, until winter tides and northeast gales forbade even the most hardy and reckless of navigators.

Yet, she thought, with a sudden flash of pride, we count for more than strangers to our ways ever stop to realize. When they sit down to their supper parties and exclaim over the lobsters they are about to eat, we are their providers. They could not do without us, the work of our hands, the order of our days. During the winter we bend the tough spruce withes to hold the laths of the traps and in the spring sink them in the waters off the dangerous ledges and in a hundred bays and backwaters. We go out at dawn to haul them, whatever the weather. It is we down here on these unknown coves and islands upon whom they unconsciously rely. Without us, what would the big hotels do in New York, and Philadelphia, and Boston? What would people eat for their grand parties? They would never see or even think about Sam Parker, or Nora Blodgett, or Ben Stevens, and hundreds like them, and yet they were dependent for their pleasures upon the toil of each.

She went indoors and upstairs to replenish her fire, for the men could not be long now in returning. They would be cold and hungry, wet from trudging over the flats now that the tide had gone. Nor could people who lived in great cities, she thought, ever know the sense of dependence upon one another which a tiny and isolated community accepted and relied upon as its daily fare. No matter how they might be reft by petty rivalries and by paltry curiosities and suspicions, they were nevertheless bound inseparably in their common labors and their common dangers.

She smiled, as she measured out the coffee, put it in the pot and broke an egg into it after her mother's recipe, in anticipation of their winter coffee parties in one fish-house or another. After the weather really closed down upon them, in December and January, they would hold these parties, men and women together, on dark mornings of high wind or snow. The fish-houses smelled of pungent spruce laths and boards, their floors powdered with sawdust, their new yellow traps piled against the walls ready for their twine fittings. The men dropped their saws and hammers, stirred up the fires in the air-tights, lighted their pipes, and enjoyed to the full the hot coffee and the freshly made doughnuts. Outside the high tides thundered upon the beach, or the icy, brown flats discouraged the swooping gulls. Inside there were warmth and cheer and even laughter as all together they forgot the long, gusty nights and the bitter, lowering dawns.

When she came down through the lighted store to return to the porch and listen for the throbs of the engine, she was again seized by pride as she looked about upon her now depleted shelves and counters. She and Joel were feeding people near at hand so that they might in their turn provide for multitudes far away. They two also were more necessary than she had ever before realized. They were the source not alone of daily bread, but of warmth and companionship, of credit through lean winters, of hope and confidence. Had they stayed on their island, where Joel could never have become the mainstay which he had made himself here, they could not have been granted the compensations which were theirs.

In point of fact, there were numberless compensations in the only life she knew. The old days, which Sarah Holt had known and which had made her what she was, had gone; but the coast remained. The great ships, which it had once built and launched and sent to the far corners of the earth and the sea, had given place to grubby fishing-boats; their captains, who wore fine, tucked shirts and dined in spacious cabins, to men in hip-boots and oilskins. Its present days, in comparison, were changeless and uneventful; yet they brought their gifts. When people said to one another after months of cold and sleet, "It's almost like spring," they were seeing with fresh pleasure the curve of the shoreline freed from ice, the sunlight on the wings of the gulls, the calm, incoming tide. On some late March or early April morning, the fishing-boats would be put back into the water amid great bustle and excitement. The martens would come back to the bird-houses, the swallows to the eaves of the fish-houses. Women who had thumbed for weeks through the seed catalogues now sent off their money-orders by the mailman. On some clear day, when the sun, sea, and air were in a certain peculiar relationship, a mirage took place, during which distant points and islands moved so close to their own cove that the whole long, ragged coastline lost its isolation and seemed curiously brought together in one accord and in one place.

She thought, too, of other small communities like their own on other far-distant shores, brought suddenly into the same accord because of the unifying force of the sea and the identity of their common labor. There were remote coves like this in Norway and in Iceland, in England and in Ireland, and on those western Scottish islands where that boy had come from who had played his jig. People there on those other shores of the world knew, as she and her neighbors knew, the long, eternal rhythms of the tides, the terror of the fogs, the wonder of the sun. Even now they were sleeping, or rising to launch their boats, or waiting for the runs of their particular kinds of fish, speaking in strange tongues perhaps, but all knowing the mysterious, uniting language of the sea.

Now with the autumn creeping on, this dark headland, this hill to the north, and Shag Island would burn with flames of color among their dark spruces. The wild geese would fly southward, their long black arrows against the sky not always bringing loneliness and dread of the winter. Even in the winter itself, when the few small houses looked unprotected and bare and the sea asserted its dominion over the shore with mighty tides and tempestuous winds, the very defiance and endurance of those who still stayed on, worked over their traps and in their kitchens, and waited, brought forth admiration and wonder that men and women were what they were, insignificant, unknown, and yet invincible in what Sarah Holt had called the circle of life. There was wonder, too, as in this hour, in these

strange returns of faith, which, unbidden and even unexpected, now and then enclosed the human spirit, shutting
out fears, redeeming one's place and time, making possible
one's future.

She heard the first distant throbbings of the engine. In
a few minutes more they would be in, sad perhaps, and
surely ready for their supper. They must now be seeing
the nearer glow of the lights in the houses. When one but
thought of last night's fog, what a miracle this day had
been!

9

Joel and Sam were quite ready for their supper. Thaddeus had not wanted to come with them, Joel said, seeing
the extra place set up on the table. Nor would he go home
with Benjamin Stevens, although Ben had been on the
beach to invite him. He wanted to be by himself, he said;
but he asked them to thank Lucy especially for the lamp
which she had left burning in the house for him.

Lucy kept filling their cups with hot, strong coffee. She
ran downstairs to fetch some more honey for their biscuits. She had placed some flowers on the table, and the
kitchen was warm and cheerful. When they had eaten and
the men had helped her to tidy up, they all went into the
sitting-room. Joel and Sam loosened their white collars,
took off their good coats, and lighted their pipes.

Now they looked as they had looked on numberless
other evenings, Lucy thought, just as she did herself in her
old house dress with her mending-basket on the table. It
might be any evening throughout any year, in spite of all
that had happened since the day had dawned, clear and

still. The words of the story which had returned to her that morning came back again. They were true. The meanings of things were forever slipping away. Even though you had thought a few hours ago that you had them for always, they escaped you, disappearing like the wings of birds into the fog or like one tide into another. Yet the loss of them did not always leave one searching and lonely, as the speaker in the story had said. She had never been less lonely than she was at this hour, nor was she, for once, seeking after anything at all.

"I somehow can't feel sad," she said, placing her darning-ball in a sock of Joel's. "Did it seem awful lonesome out there on Shag Island?"

"No," Sam Parker said. "I don't think any of us felt that it was lonesome. It was all just as it should be, to leave her there where she was born and where things once made her life what it was. I kept thinking all the time, Lucy, of what you said when you came into the store night before last, how her death was the end of an era on this coast. It couldn't have been said better."

"It takes Lucy to say things in just the proper way," Joel said. "No one except Lucy would ever have known how to put it in just them right words."

Lucy held the darning-ball in the heel of Joel's sock more tightly. It was queer, she thought, how the best time to tell certain things never seemed to come; how, if you told them at the wrong time, spells were broken, like the spell of these quiet minutes with all three of them close together in their minds and feeling the same assurance that everything was, for the moment, right.

"I was hoping all along that you got the sun out there,"

she said, "longer than we kept it here with the headland so high."

"Islands always seem to hold the sun longer," Sam said. "Perhaps that's because of the sea on every side. Anyhow, it didn't seem really dark, did it, Joel?"

"No," Joel said, "it didn't. I've always noticed on this coast how just on the edge of darkness, the sky often holds a long, steady glow of light."